INDUSTRIAL ELECTRONICS

FOR ENGINEERS, CHEMISTS,
AND TECHNICIANS,

WITH OPTIONAL LAB EXPERIMENTS

INDUSTRIAL ELECTRONICS
FOR ENGINEERS, CHEMISTS,
AND TECHNICIANS,

WITH OPTIONAL LAB EXPERIMENTS

Daniel J. Shanefield

RUTGERS UNIVERSITY

SciTech Publishing, Inc.

Distributed by
William Andrew Publishing
www.scitechpub.com

Library of Congress Catalog Card Number: 00-52188
ISBN: 0-8155-1467-0

Published in the United States of America by
Noyes Publications / William Andrew Publishing, LLC
13 Eaton Avenue
Norwich, NY 13815
1-800-932-7045
www.knovel.com

Library of Congress Cataloging-in-Publication Data

Shanefield, Daniel J., 1930-
        Industrial electronics for engineers, chemist, and technicians / by
        Daniel J. Shanefield
        p.   cm.
        Includes bibliographical references.
        ISBN 0-8155-1467-0
        1. Industrial electronics. I. Title

    TK7881 .S52 2001   2001
    621.381--dc21   00-052188
                                                                    CIP

Transferred to Digital Printing 2009

CONTENTS

PREFACE

This book can be used as a resource for working engineers and technicians, to quickly look up problems that commonly occur with industrial electronics, such as the measurement noise due to "EMI," or oscillations from "ground loops." Sufficient understanding can be obtained to solve such problems, and to avoid additional problems in the future. Most other books in this field are oriented toward electronics specialists, and they are more difficult for chemical, mechanical, or industrial engineers to use for this purpose.

The book can also be utilized as a text for a first-year laboratory course in practical electronics, either in vocational high schools, or in various college-level engineering schools, or in company training programs for people who are already in the work force. This course was designed with a view toward the fact that a great deal of electronic equipment for measurement and automation is in use nowadays, and technologists are often faced with difficulties due to misuse of equipment or failure of various components. It has been the author's experience in industrial jobs that a basic understanding of electronics can often prevent misuse, and it can aid in diagnosing equipment failure. Quite often a basic understanding can also lead to improvising new circuits that are simple but still very useful.

The experiments can be done in an ordinary classroom or conference room, without special laboratory facilities. The instructor can be anyone who has studied high school physics. Except for the oscilloscope (which might be shared by a "team"), all of the components can be purchased at Radio Shack stores or similar sources, and the equipment list has been kept to a bare minimum. In fact, the book can easily be read by itself, without experiments. In that case, the "experiments," can be considered to be examples of the circuits being explained.

The use of minimal equipment, in addition to needing less investment of money and time, has an important advantage: the function of *every* component in every circuit can be explained in the text, without taking up too much space in the book. The author has tried to use some other textbooks to teach this type of course, and there were usually a few unexplained "mystery components" in each of the complex circuits being constructed. These mysterious things did not give the students confidence for improvising their own electronic applications in future situations. Also, it limited the instructors to people who were expert enough to answer the students' questions. Therefore, the present book includes only the types of circuits where it is not necessary to optimize by means of a large number of extra components. In spite of this, it might be surprising to a knowledgeable reader that many of the most important concepts of industrial electronics are actually covered in the book, at least at a simplified but usable level. In the author's experience as a teacher, this is as much as most first year students will be able to remember, several years into the future, unless they take additional courses that repeat some of the material.

With the above comments in mind, it should be apparent that this course only *barely* touches upon the *advanced* concepts of electrical engineering. It does not provide much direct training for specialists in electronic design. However, former students have told the author that this course gave them enough information so that, when working on their new jobs, they were able to devise useful circuits, use oscilloscopes, etc., and thus solve various problems.

Some of the topics covered in the book might be difficult to find in other books, including the avoidance of measurement errors caused by excessively high or low input impedances, reading electrician's (as contrasted to electronic) symbols, understanding the shaded pole ac motor, getting 208 volts from delta or wye three-phase transformers, and optimizing a PID furnace controller.

The author has found that people are likely to remember the information for a longer time if they actually do each and every experiment with their own hands, including starting from the beginning with the oscilloscope, without much help from partners. There seems to be a hand-to-brain linkage of some kind in learning engineering subjects. Also it builds confidence to occasionally make wiring mistakes, and to learn the procedures for finding them and correcting them, without needing outside aid.

If laboratory funding is not available, a useful alternative is to use the book as a special reading assignment for an existing course, without experiments or lectures, because the book is self-explanatory. A short examination could be given, and grades might even be limited to pass or fail. Another possibility is to have the reading be done during the summer vacation period.

A teaching strategy that appears often in this book is the use of analogs. Readers almost always have a natural feeling for the way water would flow in a *wide* pipe versus the flow through a *narrow* pipe, and this is used in the book as an analogous illustration of the flow of electricity in good conductors versus resistors. Water analogs are also called upon to explain the mathematical formula for electrical resistors in parallel and other concepts throughout the book. Some of the author's students of ten years ago, including E.E. and physics majors, have recently reported that these analogs helped them achieve a deeper understanding of devices such as ZnO varistors, and therefore they still remembered the electrical behavior (V versus I diagrams) very clearly.

Modern technological jobs require an increasing amount of theoretical knowledge, and therefore many engineering colleges have been eliminating laboratory courses, in order to leave time for the teaching of more theory. Also, the vastly increased complexity of modern electronic equipment can make lab courses too expensive, unless computers are used for simulation instead of making the real, handmade, hard-wired circuits. At the same time, the old hobbies of repairing automobiles and building electronic kits that previously provided much of this experience have largely disappeared. Because of these trends, industry supervisors have begun complaining to professors that the recent graduates no longer have firsthand experience with such things as soldering or a high impedance voltmeter, let alone an oscilloscope. If they try to wire a circuit and make a mistake, they have no idea how to find this error and make their own corrections. They do not have the confidence to improvise new circuits, even simple ones, for such things as amplifying signals from sensors. Nowadays these basic skills must be learned, sometimes inefficiently for a year or more on the job, before many new employees become productive.

The author grew up in the days of do-it-yourself crystal radios and the later hi-fi stereo kits, kept pace with new developments, and in fact innovated a small amount of the new electronic technology now being used worldwide. While working for several decades in the factories and laboratories of AT&T and Lucent Technologies, he was often asked to help solve problems simply because of that previous experience. This book is an attempt to share such knowledge with a widely varying audience, in a simplified format. It is hoped that the use of this book might increase the productivity of many types of workers in science and engineering.

DANIEL J. SHANEFIELD
RUTGERS UNIVERSITY

CHAPTER 1

Introduction

Instead of an introductory chapter that presents a mass of text about the history of electronics, or its importance in modern life, this chapter will start right in with experiments illustrating the "inductive kick" that sometimes destroys expensive computers. These experiments also include making a simple radio transmitter of the type that saved 600 people on the ship *Titanic*.

THE INDUCTIVE, DESTRUCTIVE KICK

When electricity flows through a coil of wire, the physical phenomenon of "inductance" becomes strong enough to be easily detected. This is similar to a heavy iron piston moving through a water pipe, along with the water. It is difficult to get it to start moving, but once it moves, the heavy piston is hard to stop. Of course, with the heavy mass of iron, the phenomenon is commonly called inertia. This can be considered to be an "analog" of inductance, which means that, although inertia and inductance are not really the same, they behave similarly in some ways. Electricity moving through a coil (in other words, through an "inductor") is hard to start, but it is also hard to stop after it has started flowing. In fact, it is so hard to stop, that it can cause a lot of trouble if you try to stop it too quickly.

A better understanding of inductance and other features of wire coils will be provided by later chapters in this book. However, in this chapter just the behavior itself will be studied, without analyzing why it behaves this way.

1

EXPERIMENTS

Before beginning the experiments, a few procedural things have to be covered. The source of electricity will be a 9 volt battery, and the connections will be made through *clip leads*. (The latter word is pronounced "leed," not "led" like lead metal would be.) In supply catalogs, the clip leads are sometimes described by other phrases such as "test leads," "jumper cables," or "patch cords." Because of their appearance, the adjustable connectors at the ends are called "alligator clips."

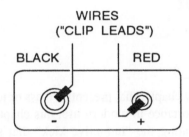

WIRES
("CLIP LEADS")

BLACK   RED

Figure 1.1 Special arrangement for attaching clips to a 9V battery.

By squeezing the large end of the alligator clip, along with its soft plastic insulator, the small end of the clip will open, and that is then placed on the rim of one circular metal terminal of the battery, and the opening force is then released. While this is quite obvious (almost insultingly so), what is not obvious to many students is that the two metal clips (positive and negative) must be carefully prevented from touching the outer metal casing of the battery, or touching each other. This can best be done by arranging the two clips as symbolized by the black rectangles in Fig. 1.1, although the wires are actually coming out of the page toward you, and not going upwards as shown in the figure. Black wires are usually put on the negative terminal and red on the positive one.

If the plastic covering slips off an alligator clip, which does often happen, open the clip as before, and then put your *other* hand "in the alligator's mouth," which can be done without hurting your fingers by squeezing the imaginary "animal's cheeks" sideways into its "mouth." Holding the clip open in that manner, your first hand can easily slip the plastic back over the large end of the clip. (Students who did not know this trick have been observed by the author to be angrily wrestling with those slippery plastic insulators, eventually giving up, and then letting the clips remain uninsulated.)

Following the circuit diagram of Fig. 1.2, run the battery current through the 120V/12V transformer, using only the "secondary" side. The way to interpret Fig. 1.2 (hopefully not being too obvious) is to attach one end of a black clip lead to the bottom end of the battery as shown in the figure. This is the negative terminal, which is the larger ("female") metal circle on the end of the actual battery, as shown previously in Fig. 1.1.

Connect the other end of that same clip lead to either one of the two secondary wires on the transformer, which both have *thin yellow* plastic insulation on them. Do not use any of the black wires of this transformer, either the thinly insulated "center tap" of the secondary coil or the two thickly insulated black wires of the "primary" coil. (This experiment can be done either with or without a long "power cord" and plug attached to the primary.)

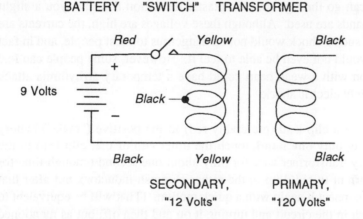

Figure 1.2 Generating a pulse by stopping the current in an inductor.

The reader probably knows from high school science courses that the *primary* coil of this transformer usually has several hundred "turns" of wire in its coil, although the transformer symbol used in this book only shows 3 turns. The *secondary* would have only one tenth as many turns, but for simplicity, each of the "windings" is shown here as having 3 turns. In this experiment the windings are not being used as a transformer — we are merely using one part as a simple inductor.

Negative wires are often considered to be "grounds," even though this one is not actually connected to the true ground. It is usually best to be consistent and have black or green colored wires be the negative ground connections, in order to avoid mistakes. It is also best to make all the ground connections first, because in case mistakes are made later (which often happens when the circuits get more complicated), it is easier to trace errors if the grounds are all completed before attaching the positive wires. The negative ground connections are arbitrarily defined to be at zero voltage, so the positive wires can be considered as being +9 volts "above the ground potential."

At the upper end of the secondary coil, the symbol that is labeled "switch" in the diagram represents a contact that is made and then broken, repeatedly. It could be a real switch, such as you would use to turn on the lights in a classroom, but to save money we will just use one end of a clip lead that is touched for *a short time* to the upper transformer secondary wire.

SAFETY NOTE: Do not touch wires with more than **one hand** at a time while generating an inductive kick in the next part of this experiment. The high voltage can go through the thin plastic insulation and give you a slight shock if two hands are used. Although these voltages are high, the currents are very small, so such a shock would not be dangerous to most people, and in fact most people would not even be able to feel it. However, some people can feel it, and a person with a weak heart could have a temporary arrhythmia attack with even a slight electric shock.

Now connect a clip lead (preferably red) to the positive ("male") battery terminal. Using only **one hand,** touch the other end of that clip lead to the other secondary transformer wire for only about one second (enough time for the electric current to build up in the rather sluggish inductor), but after that short time, disconnect it again with a quick motion. (That will be equivalent to having a switch in the circuit and turning it on and then off, but as mentioned above, an actual switch will not be used.) A small spark will appear when you disconnect the wire, because the electricity has a strong tendency to continue flowing — the inductance of the coil causes it to behave this way. Instead of suddenly stopping, the electricity builds up enough voltage (in other words, enough driving force) to continue flowing in the visible form of a spark, going right through the air. But as your hand quickly moves the alligator clip farther away from the transformer wire, the distance soon gets to be too great for the available voltage to continue pushing the electricity, so it stops. Thus the spark only lasts for about a thousandth of a second.

There was no spark when you made initial contact, only when you broke the contact. However, you might have caused the two pieces of metal to "bounce" (make and then break contact very fast) before settling down, while you were trying to push them together. In that case there would be a visible spark when you "made contact," because you were really making and then breaking it, and the breaking action was where the spark actually occurred. This is referred to in electronics as "contact bounce," and although we try to avoid it, various switches, push buttons, and computer keyboards do sometimes have contact bounce, and it can cause errors in computerized data. Circuits for preventing the effects of this will be discussed in later chapters.

If the sparking is repeated many times, the battery will be temporarily drained, and the spark might stop appearing. In that case, wait a minute for the battery to recover its proper voltage and then try again.

This spark is not very impressive. However, the little spark represents a very high voltage, even if it only exists for a small time. In fact, if you tried to use a voltmeter or oscilloscope to measure this high voltage, it might destroy those instruments. Instead, we will use a "neon tester bulb" to get an estimate of the voltage, where even an extremely high voltage will not damage it.

On the wall of your classroom, find a 120 volt electric socket. Into the two rectangular holes of that socket, plug in the two wires of the neon tester, being **very careful** not to let your hand slip forward and touch the metal "lugs" at the ends of the wires. Exert your will power to **use only one hand** for this operation, resisting the urge to make things easier by putting both hands on the wires. (If you happen to slip, the shock of having the 120 volts go from one *finger* to another, all on the same hand, would not be as likely to kill you as having it go from one *hand* to the other, across your heart area. Therefore, get in the habit of only using one hand when working with any source of more than 50 volts. This is the electrician's "keep the other hand in your pocket" safety rule.) If you are *not* in the United States of America, possibly you will need to plug into differently shaped holes (round or L-shaped), and you might have a different voltage, but the experiment will be similar.

Some instructors may insist that students wear rubber or cloth gloves during this part of the experiment, or possibly the instructor will be the only person allowed to do it. In case someone is apparently becoming paralyzed by accidental contact to the high voltage source, do **not** help that person by grabbing the body with your bare hands, because you might also become shocked and temporarily paralyzed. Instead, either use a gloved hand or else push the person away from the wall socket with your foot, only making contact with a rubber soled shoe. Although kicking your friend when he is paralyzed sounds humorous, of course an electric shock is not funny when it occurs. It is important for students to realize that **NO PRACTICAL JOKES** are allowed in an electronics laboratory, not even just false "Watch out" warnings.

Although *careless* behavior can be deadly, *careful* behavior has prevented the author from ever getting even one strong electric shock in many decades of experimentation. Working with electricity is potentially dangerous, but like driving a car or a bus, it can be done safely.

When plugged into the wall socket, the neon tester lights up, just as the reader probably expected. It requires a high voltage (more than 90 volts) but very little current (only about a millionth of an ampere). To see how little current is necessary, carefully plug only *one* wire of the neon tester into the smaller rectangular hole of the wall socket (in the U.S.A.), and grasp the *other* metal lug of the neon tester *with one hand*. Do not let your skin touch any random metal such as the face plate of the wall socket, or a metal table. (Possibly the instructor will want to do this experiment, instead of having each student do it.) The neon bulb will light up, though only faintly. A piece of paper or cardboard may have to be held around the bulb to keep the room light from obscuring the faint glow. Also, the first wire may have to be plugged into a different hole, if the socket has been wired improperly, which sometimes happens. In most cases, enough electricity goes through your body to make a visible glow, even if you are wearing rubber soled shoes. The alternating current ("ac") capacitance of your body is involved in the conduction of the electricity, in addition to the direct current ("dc") resistance, but full explanations of these concepts will have to wait until later, step by step (page 90). Suffice it to say that the neon tester responds to high voltage, even though the current is extremely low, less than you can feel as an electric shock.

If only a *single* wire of the neon tester is plugged into either of the *other* two holes in the wall socket (the larger rectangular hole or the round one), the bulb will not light when you touch the second neon tester wire. The reason is that you are acting as a return path ("ground") for the electricity, and those other two socket connections are also grounds, although they are not exactly the same types. Further explanation will be developed, as we go along.

Now attach both wires of the neon tester to both terminals of the 9 volt battery, either with or without clip leads. (Of course, this is not a dangerous part of the experiment, since 9 volts is very unlikely to hurt anyone with reasonably dry skin, but it still makes good sense to avoid any two-handed contact of metal parts having voltage on them.) The neon bulb does not light up, even faintly.

The Neon Flash
The next step is to attach the neon tester bulb to the inductor, as shown in

Fig. 1.3, using clip leads. The black dot symbol inside the circular bulb symbol means that there is some neon gas inside the bulb, and not an extreme sort of vacuum. A different black dot symbol occurs at the places where three wires come together, and this means that the wires are mutually connected together electrically. (This is in contrast to a possible situation, which you will see later, in more complex diagrams, where two wires cross but are not making electrical contact, because of insulation layers — no black dots are used then.)

It would be a good idea to use a red clip lead from the positive terminal of the battery to the neon tester lug. Then one end of a yellow clip lead is also attached to that same neon tester lug, but the other end of the yellow lead is not hooked up to anything yet — it is going to be the equivalent of the switch. A white clip lead can go from the *other* lug of the neon tester to the upper yellow wire of the secondary coil (that is, "upper" as drawn in the diagram, but not really up or down). The bare metal tip of that same upper secondary wire can then be momentarily touched by the unattached end of the yellow clip lead.

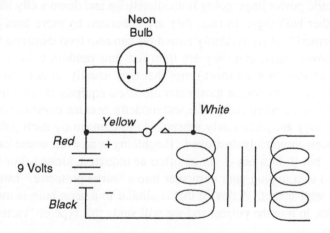

Figure 1.3 Using the inductive kick to light a neon bulb.

(In electrical terminology, when the metal wire is touched to the metal end of the clip lead, contact is "made," or we could say that the makeshift switch is "closed." Later, when the metal parts are moved away from each other, we could say that contact is "broken," or the switch has been "opened.")

When initial contact is made by this makeshift switch, nothing visible happens. But when the contact is broken, the neon bulb lights up for an instant, similar to the spark experiment described previously. Therefore the inductor has generated more than 90 volts, from only a 9 volt battery. At this point in the course, the explanation of the high voltage will be limited to a simplified one: the "inductance" of the coil causes the electricity to have a tendency to continue flowing, and since the contact is broken, the electricity can only go through the neon bulb or else stop. The fact that it does light the bulb means that we have been able to estimate the strength of this tendency to continue flowing, and the tendency to continue amounts to "at least 90 volts of inductive kick." The faster the *breaking* of contact, the brighter the light, although this might not be easily visible.

With these experiments, it was demonstrated that an inductive kick can turn 9 volts into more than 90 volts, even though the inductor is only a small coil. Imagine what happens when the large coil of a 480 volt electric motor is suddenly turned off, while it is running the air conditioning system of an office building. This does happen, every time the temperature inside the building reaches the desired value, and the big coils of the motor generate very large inductive kicks. These high voltage "pulses" or "spikes" sometimes travel back into the electric power lines, going immediately up and down a city block and into many other buildings. In fact, they are generated by more than just motors. Transformers that are suddenly turned off can also feed thousand volt pulses into the power lines, and they are then fed into random offices and houses. The pulses last for such short times that they usually do not damage other motors, light bulbs, radios, or most older electrical equipment. However, during the last few years, when computers and modems became commonplace, it was found that they are particularly susceptible to damage by such pulses, and many have been completely destroyed. (Lightning is an even worse cause of pulse damage, but it does not occur as often as inductive kicks.) For this reason it is a good idea to plug your computer into a "surge protector," instead of directly into a wall socket. A device that is similar to a neon bulb is inside the surge protector, to trap the pulses, and we will study this type of "varistor" later in the course.

Repeat the above experiments, but use the *heavily insulated black* wires of the *primary* coil in the transformer instead of the yellow secondary wires. The neon bulb flashes slightly brighter, because the voltage generated is higher. However, the spark is much dimmer, because there is less flow of electric current during that fraction of a second. In the experiment on the next page, the effect will be stronger with the primary than if you had kept the secondary hookup, because voltage is the important factor, not current.

The Radio Transmitter Of The Ship *Titanic* (Simplified)

If a portable AM/FM radio receiver is available, place it about three feet from the inductor and battery. (If a small radio is not available, the inductor and battery can be taken outside, near an automobile radio.) Turn on the radio, set it for AM, and tune the frequency dial to a number where there are no nearby radio stations, so only a very low level of background noise is audible. Turn the volume up somewhat. Disconnect the neon tester, and then make repeated sparks with just the inductor and battery, as shown in Fig. 1.2, but using the primary coil (heavily insulated black wires). Loud clicks will be heard from the radio loudspeaker, each time a spark is made. The high voltage pulse generates a radio wave, which travels through the air to the radio antenna. (Further analysis of these various aspects of radio theory will be explained in later chapters.)

AM ("amplitude modulation") is sensitive to this type of "radio frequency interference," or RFI. There are so many different kinds of electrical machines generating inductive-kick RFI, that AM radio reception in a crowded city environment is plagued by background noise, which is usually called "static." Set the radio to FM ("frequency modulation") and repeat the experiment. FM is far less sensitive to RFI, and this is one of its biggest advantages over AM.

While radio waves generated by inductive sparks are undesirable nowadays, it is apparent that they are easy to make, even with this simple apparatus. Therefore it should not be surprising that the first commercial radio system built by Guglielmo Marconi in 1912, sending pulsed Morse code signals across the Atlantic Ocean, was very similar to the one constructed in this experiment, although it was much larger. It involved an inductive kick and a spark, but it also had some other features such as an automatically repeating switch, a long antenna, and a tuning system, and those will be described later in this book. A nickname commonly given to radio operators in those days was "Sparks."

One of the first practical radios was a Marconi-designed spark transmitter on the steamship *Titanic,* which was used by a heroic radio operator to attract the other ship that saved the survivors. After that highly newsworthy accident, the use of radio increased rapidly.

GRADING BY THE INSTRUCTOR

A "pass or fail" grading system might be appropriate for this course, based on attendance and successful completion of the experiments. However, if ordinary grades are required, they might be based on some combination of attendance, open-book lab reports, and closed-book examinations requiring the students to draw the circuit diagrams.

ADDITIONAL READING

Much more can be learned about many of the topics in this book by looking them up in the index of *The ARRL Handbook,* edited by R. Dean Straw (American Radio Relay League, 1999). The publisher is in Newington, CT 06111 (http:// www.arrl.org).

EQUIPMENT NOTES

Components For This Chapter	Radio Shack * Catalog Number
Battery, nine volts	23-653 or 23-553
Clip leads, 14 inch, one set.	278-1156C or 278-1157
Neon test light, 90 volts	22-102 or 272-707
Transformer, 120V to 12V, 450 ma, center tapped	273-1365A or 273-1352
Portable AM/FM radio	12-794 or 12-799

When putting away the components until the next lab session, fold each clip lead in half, and then line them all up next to each other, in parallel, with the folded parts together at the top. Place a rubber band or short wire twist around this group of clip leads. Put the battery in a small, electrically insulating plastic or paper bag.

---

* These and similar items can be purchased from the Radio Shack mail order subsidiary, RadioShack.com, P. O. Box 1981, Fort Worth, TX 76101-1981, phone 1(800) 442-7221 , e-mail commsales@radioshack.com, website www.radioshack.com.

CHAPTER 2

Ohm's Law and Measurements

THE WATER ANALOG

Imagine a water pump system as shown in Fig. 2.1, with a handle (at the left) attached to a piston, being pushed toward the right-hand side. The water then comes out of a faucet, into an open pan, and one-way check valves prevent it from going in the wrong direction. When the handle is pulled back toward the left side, the water is sucked up into the pump again.

Readers will intuitively have a good understanding of how this system would behave. That is, pushing harder on the handle would raise the pressure,

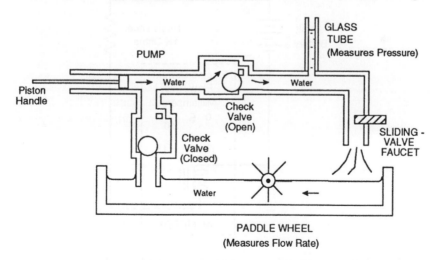

Fig. 2.1 A water pump "analog" of a simple electronic system.

11

and therefore the level of water would rise visibly in the glass tube. As the *pressure is increased,* the *flow rate of water increases,* as measured by the speed of the paddle wheel.

Alternatively, a different way to *increase the flow rate* would be to open the faucet more, by sliding the cross-hatched part of the valve toward the right. This *decreases the resistance* to flow.

The effects on flow rate as described above might be described mathematically by equation 2.1.

$$\text{Water flow rate, in liters/minute} = \frac{\text{Pressure, in grams/square cm}}{\text{Resistance to flow (no standard units)}} \qquad (2.1)$$

This equation assumes that the relationships are linear, and they actually are with water, as long as the flow rate is low.

The reason for bothering with all these pumps and valves is that the very familiar behavior of water is similar to the less familiar behavior of electricity. This similarity will aid in understanding electronics later in the course, when things get more complex. Electricity, shown in the diagram of Fig. 2.2, is said to be "analogous" to water, because the two behaviors are similar in some important ways.

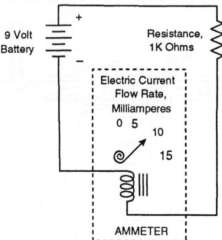

Fig. 2.2   The electronic circuit analogous to the system of Fig. 2.1.

Nine volts from the "battery" in the diagram is analogous to pressure from a water pump, although they are only similar, and not really the same. The electrical resistance in this case could be provided by a very *narrow* wire (a "resistor"), instead of a *narrow* opening in a pipe. (See also page 265 of Appendix.) There are other ways to make resistance, and the zig-zag symbol does not indicate which method is being used, only that some resistance is there.

The flow rate of electricity is measured by the strength of a magnetic field, coming from a stationary coil of wire at the bottom of Fig. 2.2, and pulling down on a movable iron pointer that is at the end of a spiral spring. The three vertical lines next to the coil are a symbol for several sheets of iron, which concentrate the magnetic field towards the pointer. The more current flowing through the "electromagnet" coil, the more the pointer is pulled down toward the stationary magnet, thus pointing to a higher number of "amperes" (units of current).

Taken all together, the coil, the iron sheets, and the spring-pointer-numbers assembly (the things inside the dotted line) are referred to as an "ammeter." This ammeter is the most common type and will be used throughout the course.

OHM'S LAW

The various effects on the flow rate of electricity in Fig. 2.2 can be described mathematically by equation 2.2.

$$\text{Current, in amperes (or coul./sec)} = \frac{\text{Voltage (or "potential"), in volts}}{\text{Resistance, in ohms}} \qquad (2.2)$$

As many readers already know, this is "Ohm's law." The current is driven forward by the voltage. The "ampere" unit is the same thing as "coulombs per second," where one coulomb is 6×10^{18} electrons, so it is a *quantity* of charge *moving per* unit of *time*, or a flow rate. In many ways equation 2.2 is analogous to equation 2.1. In fact, if the reader ever has trouble remembering which term is the numerator in Ohm's law and which is the denominator, it will be easy to recall that a water flow rate is increased when the pressure increases, and therefore the pressure is on top of the fraction, just as voltage must be.

In standard symbols, Ohm's law is

$$I = \frac{V}{R} \qquad (2.3)$$

At one time, the "I" stood for the word "intensity," but that word is no longer used to mean electric current, and only the first letter remains, preserved as part of Ohm's law.

In Fig. 2.2, if R = 1,000 ohms, then I = 0.009 amperes, or 9 milliamperes. Common **abbreviations** for these values are "1 K" (which is short for 1 kilo-ohm) and "9 ma."

If R is *not* known, but the current and voltage *are* known, then R can be calculated by rearranging the equation to be

$$R = \frac{V}{I} \tag{2.4}$$

or V can be obtained from the other two by the equation

$$V = IR \tag{2.5}$$

which is the form that some people use for memorizing the relationship.

In books written a few years ago, the terms "potential" or "potential difference" were often used to mean voltage, but they are only rarely used now. Also appearing in older books was the symbol E, which was almost always used for voltage instead of using V. The letter E is a shortened form of "EMF," meaning "electromotive force," but voltage is not really a force, so these terms do not appear often in modern publications. (The new meaning for the term "EMF" is on page 57.)

FORCE

Forces of *attraction* are present when a group of electrons (negatively charged) is brought close to a group of positive charges, but forces of *repulsion* would result if *both* of the groups being brought together are electrons. These are "electrostatic" forces, which are somewhat analogous to magnetic forces. If two groups (call them "a" and "b") of electrons are spread out on flat planes or on spheres or other shapes, the repulsion *force* can be calculated as

$$\text{Force} = k \frac{(\text{Coulombs }_a)(\text{Coulombs }_b)}{(\text{separation distance})^2} \tag{2.6}$$

where k has slightly different values, depending on which type of geometry is used. The number of coulombs involved is the electric "charge," and the dimensions of this **force** are **(charge)² / (distance)²**.

[Going back to water pipes for a moment, the usable **force** can be calculated by multiplying water **pressure** times the **area** of a piston on which the pressure is operating. We will be comparing this to electricity, within the brackets that appear toward the bottom of this page.]

ENERGY

The electrostatic force continuing to act through a slowly changing distance can be used to calculate *work* (or *energy*), which is the calculus integral of *force times* the increments of *distance*. Since "distance" is multiplied against those "force" units in equation 2.6, one of the dimensions of distance cancels out, so the dimensions of **energy** become **(charge)^2 / distance**.

FIELD

Imagine a large, stationary metal plate that has many excess electrons spread over its surface, and therefore it has a strong negative charge. It is useful to be able to measure the *electric field* around this charged object. A standardized way to do that is to bring a small movable object (possibly a tiny metal plate) nearby, with only a very small electric charge on it (possibly a single excess electron). There will be a repulsion force as described in equation 2.6. Then the *electric field* is

"*force* per unit *charge* on a small movable object."

Its dimensions are *force divided by* the *charge* on the small object (called the "test charge"). Therefore one of the charge dimensions cancels out of equation 2.6, and the remaining dimensions of **field** are **charge / (distance)^2**.

[For readers who are curious to make comparisons, the usable **force** can be calculated by multiplying electric **field** times the **charge** on the small movable object. This will be mentioned again, in brackets on next page.]

VOLTAGE, THE STRANGE ONE

Although we would like voltage to be similar to pressure or maybe force, which we can feel directly and therefore imagine, unfortunately the dimensions of voltage are "none of the above." **Voltage** is simply **charge / distance**.

A question that may occur to readers is "why is voltage used so much in electronics calculations, when it is *not* some kind of *pressure* or *force?*" There are two important reasons. *One* reason is that *it is easy to measure voltage.* For example, the simple apparatus in Fig. 2.2 can be used, with an ammeter and a known, standardized resistor.

By comparison, if we had chosen to measure electric *field* instead of voltage, it turns out that the *forces* of repulsion or attraction are very small for fixed charges. However, if we choose to measure voltage, we can pass a fairly large amount of current through the ammeter coil in Fig. 2.2 (the current could be billions of electrons per second), and with many "turns" of wire in the coil to further multiply the effect (by as much as a factor of 100, if 100 turns are actually used), the *magnetic* force of attraction to the iron pointer can be quite large. Therefore voltage is easy to measure accurately.

There is *another* reason why voltage is an important parameter in electronics: since it is easy to measure voltage, then it becomes easy to calculate "power," and that will become apparent when reading the next section (equation 2.8). In order to understand the relationship, some further looks at the dimensions are necessary.

If the dimensions of *energy* as noted on the previous page are *divided again by charge,* the result is charge / distance, with nothing being squared. This has the same dimensions as voltage. Therefore another way to describe **voltage** is

energy per charge.

(We will make use of the term *energy per charge* on the middle of the next page, in equation 2.7.) Following are some other interesting analogies.

> Now consider the *energy* of a **spring**, while it is being compressed through a certain *distance*. For each additional little bit of distance moved, the energy stored (see "Energy" on page 15) increases by *force times* that change in *distance*. Thus, **energy *per distance*** has the dimensions of **force**.
>
> Now, consider the *energy* of a storage **battery**, while it is being charged up. Note that the **energy *per charge*** has the dimensions of **voltage,** as in the bold print above, when each little bit of charge is going into the battery.
>
> So *voltage*, which is *energy per charge*, makes electrons move, somewhat like *force*, which is *energy per distance*, making a mass of material move. (Obviously, voltage and force are only analogs, *not the same* as each other.)

MORE DIMENSIONS, LEADING TO POWER

The reader might remember from general science courses that *power* is

"energy dissipated *per time,"*

and the unit of power is the watt.

OPTIONAL SECTION

Looking again at the statement about voltage on the previous page, one definition of voltage is

$$V = \frac{energy}{charge}.$$

The next step is to recall from the bottom part of page 13 that

$$I = \frac{charge}{time}.$$

If we seek to arrive at the expression for power as

$$P = \frac{energy}{time},$$

then *voltage* can be multiplied times *current*, resulting in cancellation of charge, and equation 2.7 is the result.

$$V \times I = \frac{energy}{charge} \times \frac{charge}{time} = \frac{energy}{time} = P \qquad (2.7)$$

Using voltage, it turns out to be *easier* to calculate the power of *electrical* devices than it is for most *mechanical* devices, and the simple formula is

$$P = VI \qquad (2.8)$$

If the current is not known (but V and R are known), then Ohm's law can be used to substitute V/R instead of I, as in equation 2.9.

$$P = V^2 / R \qquad (2.9)$$

Or, if the voltage is not known (but I and R are known), then Ohm's law can be used to substitute IR instead of the V in equation 2.8, resulting in equation 2.10.

$$P = I^2 R \qquad (2.10)$$

NONLINEAR RESISTANCES

Most resistors that are used in electronics follow Ohm's law in a linear manner. According to equation 2.4 on page 14, the resistance is R = V/I. Looking at the left-hand graph of Fig. 2.3 below, the slope (signified by the curved dotted line) is V/I, so this slope is the resistance, R. In this particular case it is 2 ohms. The "characteristic curve" of an ordinary resistor is a *straight line* ("linear"), so doubling the voltage would double the current.

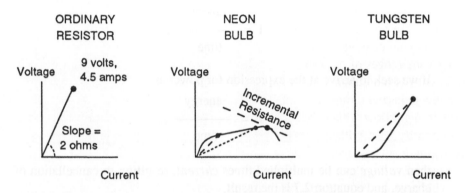

Fig. 2.3 The "characteristic curves" of linear and nonlinear resistors.

The reader might be puzzled that the "cause" of the action in electricity, that is, the voltage, is plotted on the vertical axis of the graph instead of the horizontal axis that would be more commonly chosen in algebra books. One way to reconcile this seemingly backwards way of graphing V and I is to imagine the water pump in Fig. 2.1 as being operated by a stubborn person who insists on moving the piston at a speed such that the flow rate is a certain amount. Then he changes this rate, and you are supposed to measure the pressure at each of his flow rates, but you can not control them. If you make a graph of the pressures that occur, you will obtain the water analog of Fig. 2.3, where the flow rate is the horizontal, "independent" variable (cause) and the pressure is the vertical, "dependent" variable (result). A plumber might call your measurements "back pressures," where the more the water is forced through the constant valve opening, the more pressure is developed, as an undesirable side effect of that stubborn guy increasing the flow rates.

Why do electronics researchers often plot the data this way? One reason is that the slope can be directly interpreted as a resistance. Data used by mechanical engineers and civil engineers are also plotted this way in many cases ("stresses" on the vertical axis, versus "strain rates" on the horizontal). Also, the viscosities of fluids are plotted this way.

Several important types of electrical devices such as neon bulbs have resistance characteristics that are not linear, and an example is shown in the middle graph of Fig. 2.3. In this case the resistance (R = V/I) is *not* the slope of the *solid* curved line. Instead, it is the slope of the *dotted* straight line drawn from the origin to *a particular point* on that solid curved line. At low voltage, the resistance is high (slope of the *short dashed* straight line), but as the voltage increases, the resistance goes down (slope of the dotted line).

There are no numbers given for the voltages in this simplified diagram, and with this kind of device it would be best to use logarithmic scales, to be able to cover wide ranges of numbers. In the bulb used in a "neon tester," the resistance at 80 volts is more than 1 million ohms, but above 90 volts it goes down sharply to a few thousand ohms, so much more current begins to flow. In fact, too much current might flow with a bulb hooked up all by itself, so a "protective resistance" is ordinarily used with it, making the total resistance high enough to prevent a damaging level of current. The neon bulb used in these experiments has such a resistor inside the plastic housing (not visible from outside).

This neon bulb's characteristic curve is sometimes referred to as being "non-ohmic." That term is not a very clear description, because at any given point on the curve, Ohm's law still applies, and R = V/I. A more descriptive statement would be that *the resistance is not constant.* At a high enough voltage, the neon gas begins to ionize more and more (an "avalanche" effect), and the resistance goes down.

In the region of the curve where the resistance goes down, it is described as having "negative incremental resistance." In fact, it is sometimes called just "negative resistance," but that is an incorrect term, because the *resistance* itself (the dotted line) is never negative. Superconductors can have *zero* resistance, but *negative* resistance would mean that energy could be generated from nothingness, which human beings have not yet learned how to do.

What really is negative is the "*incremental resistance,*" and that is the slope of the *long dashed* line shown in the middle diagram. While that can be negative, the *dotted* line slope from the origin (the true *"resistance"*) would always be positive.

Gas filled tubes such as some types of neon bulbs (and also some silicon diode devices) have characteristic curves that are "flatter" and more horizontal along the top than the one shown here. Examples of these will be studied later in the course. Those devices can be used to ensure that the voltage remains constant, over a wide range of currents.

Another type of nonlinear resistor is the ordinary tungsten incandescent light bulb, with a characteristic curve as shown at the right-hand side of the figure. The resistance is very low when the wire is cold, but it rises as current heats the wire more, because there is more "scattering" of the electrons.

THE MULTIMETER

An ammeter was shown in Fig. 2.2, with a separate battery and resistor attached to it. If all three are in a single plastic housing, a "multimeter" results, and an electrical diagram of such an instrument is shown in Fig. 2..4.

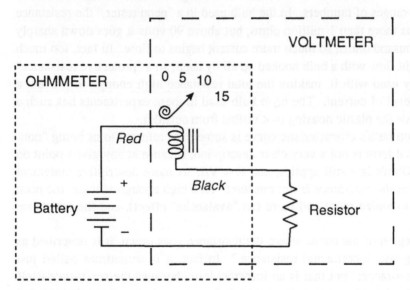

Fig. 2.4 A "multimeter."

The "ammeter" part of the multimeter is not labeled in this diagram, but it is the same assembly as appeared in Fig. 2.2 on page 12. This consists of a coil (three "turns" of wire are shown, although actually it would have many more turns), an iron core (symbolized here by three parallel lines), a spiral spring and pointer, and a stationary numerical scale. The other components (the battery and resistor), outside of the ammeter, have been rearranged, compared to the way they were shown in Fig. 2.2, but they are the same components. A small, "fast-blow" (or "fast-action") fuse is usually put in series with the ammeter for protection, although that is not shown in this diagram.

The "voltmeter" part of the multimeter is surrounded by the long-dash line in the diagram. It includes the ammeter and a standardized resistor. With the aid of Ohm's law, these two components can be used to measure an unknown voltage such as that of the battery. What is being measured is whatever is outside of the dashed line. Experiments to illustrate this concept will be performed in the next section.

The "ohmmeter" part of the multimeter is surrounded by the dotted line, and it includes a standardized battery plus the ammeter, but not the resistor. Again, what is being measured is whatever part is outside of the dotted line.

Switches inside the instrument (not shown in the diagram) permit a single function to be chosen, and various components can be either included or not included in the circuit. Other switches also are arranged to permit the use of various different resistors, depending on what range of voltage is to be measured. (Various ranges of current can be measured by the ammeter alone, but the explanation of this will be delayed until a later chapter.)

EXPERIMENTS

In this course, the multimeter is an inexpensive unit (Radio Shack catalog number 22-218 or the equivalent from another supplier, sometimes referred to as a "multitester" or "voltammeter"). Care must be used to prevent dropping the meter on a hard surface such as a table, even from a height of a few inches, because the delicate mechanism is likely to be damaged.

Plug in the small pin on the end of the black wire, at the lower left corner of the meter, near the negative ("-") symbol. On some meters the negative socket is labeled "common," and that word is assumed to indicate "negative." Of course, the red wire is then plugged into the positive ("+") socket.

Ammeter
Following the diagram of Fig. 2.4, hook up the same 9 volt battery as in the previous chapter, with a black or green clip lead as the negative wire. For the unknown resistance, use the same transformer, attaching to one of its heavily insulated black wires (the primary coil). Then attach a yellow or white clip lead to the other black primary wire and also to the long "probe" pin at the end of the black (negative, or common) wire of the multimeter.

Set the multimeter controls (rotary switch or pushbuttons, depending on the particular instrument) to the **direct current (dc)** milliampere (ma) position, where the maximum amount of current is greater than what is expected in the experiment. If the Radio Shack meter, Catalog Number 22-218 is used, this will be the "150 mA DC" setting, since only about 50 ma is expected.

Carefully connect a red clip lead from the positive battery terminal to the red wire of the meter. The pointer will probably move a little bit past the lower black number "5." The black letters to the right read "DC V mA," which means that the lower black scale is used for direct current, either volts or milliamperes.

Number scales on the meter might be confusing to new users. The highest amount of current allowed at this setting is 150 ma, and this has to be the lower scale that is abbreviated with the black "15," since that contains the digits 1 and 5, and there is no other scale on the meter that might signify that the maximum is 150. If the pointer moves to 10, then 100 ma is flowing, but 5 would indicate 50 ma.

(If *voltage* were being measured, the white numbers at the left of the rotary switch show that the 15 could stand for *either* 15 volts or 150 volts, depending on which setting of the rotary switch is used. But now we are measuring *current*, and 150 ma is the only thing that could be meant by the black number 15 on the scale.)

The needle pointer will move to approximately 60 ma, which will appear to be slightly higher than "5." The small red marks are placed proportionally to any values between 5 and 10 (or zero to 5).

The red scale is marked "**AC 15V**," and this could be used for 15 volts of **alternating current** (maximum), if the rotary dial had selected that range. The topmost black scale is for 1000 volts max., either **dc** or **ac**, depending on the rotary dial position. A moderate amount of logical thinking is required to interpret these and other meter scales, and the abbreviations are not always well planned.

Based on Ohm's law, 60 milliamperes (0.060 amperes) and 9 volts indicates that the resistance of the primary coil appears to be about 150 ohms. (When these batteries are new, they actually produce about 9.5 volts, and the voltage and resistance values will be checked in later experiments.) The *dc* resistance of the transformer is essentially linear, so when lower currents are to be measured later, it will still have about the same resistance. However, the resistance to the flow of *ac* current (the "impedance") is much higher, as will be seen in the later chapter on inductors.

The *secondary* wires of the transformer could be used for experiments of this type, but the higher current due to the lower resistance might "blow out" the fuse. (A more expensive meter could handle those higher currents.)

The milliampere settings make the meter especially vulnerable to being damaged by excessive current, since there is no internal resistor being used to act as a limiter. If a battery is directly attached to the meter by mistake, with no external resistance such as the transformer, the fuse will blow out, and the meter will no longer operate. The back cover can then be taken off, usually with a small Phillips head screwdriver, and in this case the fuse could be replaced with a 315 ma fast-blow type.

One of the most common ways that fuses are blown is to leave the meter on the "ma" setting after use, and then to hook it up directly for measuring the voltage of a battery, without rotating the dial from milliamperes to volts until *after* attaching to a voltage source. Without a protective resistance, there will be too much current. In order to prevent this from happening, the rotary switch should always be set to "OFF" *before* making the next measurement. Then, if a voltage measurement is to be made next, the experimenter *has to* rotate the dial to "voltage" in order to get a reading and will not forget to do this.

It should be noted that, in this experiment (Fig. 2.4), the black wire of the meter is not attached directly to the negative of the battery. However, it is *relatively* negative, compared to the red wire, and this relationship is all that is necessary for proper readings.

Instead of the transformer as an experimental resistance, hook up the tungsten incandescent light bulb. Even though the bulb is designed for use at 12 volts, the light is still visible with only 9 volts. Many electronic components will operate in a usable fashion at somewhat lower or higher voltages than the ideal values.

The current (and thus the calculated resistance) will be similar to the values observed with the transformer primary. However, the bulb is not a linear resistor, and at *lower* currents its resistance would be different. This fact will be illustrated in the ohmmeter experiment.

A mirror under the pointer of some meters is intended to prevent "parallax." If the experimenter looks at the pointer from an angle instead of looking straight down perpendicular to the scale, a wrong value might be observed, which would be caused by a parallax error. Ideally, the experimenter's eye should be moved until the reflection of the pointer is exactly behind the pointer itself, and thus the reflection should never be visible as a separate line.

Voltmeter

Knowing in advance that the transformer has a resistance of, say, 150 ohms, and measuring a current of 60 ma, the voltage of the battery could have been calculated using Ohm's law, if it had been unknown. The transformer is not exactly a 150 ohm device, and a more accurate value of resistance is contained inside the multimeter. Access can be gotten by setting the rotary switch to DCV 15. After doing that, touch the black and red probes of the multimeter to the battery terminals. A new battery should give a reading of about 9.5 volts. If the wires are reversed, the pointer will move a small amount in the wrong direction.

Ohmmeter

Again, knowing in advance that the battery supplies 9 volts, a resistance can be measured with the ammeter and Ohm's law. A more accurate way is to use the internal calibration capability of an ohmmeter. Set the rotary switch to "RX 1KΩ ," and then touch the red and black wire probes together. The pointer should move quickly to the right. If it does not move, take off the back cover of the meter and install an appropriate battery (a 1.5 volt AA size battery, if the meter is the Radio Shack catalog number 22-218 type). When a battery is installed, hold the long metal probes together firmly and slowly rotate the red disk that sticks out of the left side of the meter, until the pointer reads zero "K OHMS" on the red scale at the top. This will calibrate the meter, even if the 1.5 volt battery is slightly weak, or if the internal standard resistor is slightly off-value because of temperature or aging effects. (Note that the symbol for ohms, just above the rotary switch, is the Greek letter omega, Ω.)

Now touch the probes firmly to the primary wires of the transformer, or use clip leads. It will be difficult to get accurate readings with an inexpensive meter, since the red ohms scale is highly compressed, and the red number "one" indicates 1,000 ohms. The small red marks are proportional, and about 200 ohms should be indicated.

If the resistance of the tungsten bulb is measured with the ohmmeter, the available "output" of the meter is only 1.5 volts. This will not cause a high enough current to strongly heat the tungsten wire ("filament") and produce visible light. Comparing to the ammeter experiment on page 23, where the nine volt battery was able to heat the filament to incandescence, the resistance was about 200 ohms in that experiment. Now, with the filament at a much lower temperature, the resistance is only about 20 ohms, although it can be only roughly estimated with this meter.

An ohmmeter can be used as a "continuity tester," to determine whether or not wires are properly connected. More expensive meters often have a buzzer that can be used to indicate good connections (low resistances) without making the experimenter look away from the wires being tested.

The *output* of an ohmmeter does not necessarily have the same "polarity" as the red and black wires would usually symbolize for *inputs*. Looking at Fig. 2.4, it can be seen that the black wire should indeed have a positive output, relative to the other wire, when an external resistance is being measured. This can be shown experimentally by hooking up your ohmmeter to a neighboring experimenter's meter which has been set to be a voltmeter. It is not always true with other types of instruments, but it is often the case with inexpensive multimeters.

COMMENTS

By now, the reader has probably noticed that the sources of voltage (or "signals" of any kind) usually come in at the left of electronic circuit diagrams and go toward the right. There can be exceptions to this convention, but they are rare.

The calculations in this chapter, and many more electronic calculations, can be done by computer with the program "SPICE" (Simulation Program with IC Emphasis), from Microsim Corp., Irvine, CA (www.microsim.com). Also, a low cost demonstration version is available from Intusoft, in San Pedro, CA (www.intusoft.com).

EQUIPMENT NOTES

Components For This Chapter	Radio Shack Catalog Number
Battery, nine volts	23-653
Battery, AA, 1.5 volts	23-872 or 23-882
Clip leads, 14 inch, one set.	278-1156C
Fuse, fast-action, 315 ma, 5 x 20mm	270-1046
Phillips head screwdriver, for replacing fuse	64-1950 or 64-1901
Lamp bulb, tungsten, 12 volt dc, green or blue	272-337A
Multimeter (or "multitester")	22-218
Transformer, 120V to 12V, 450 ma, center tapped	273-1365A

When putting away the components until the next lab session, be sure that the multimeter is returned to the "off" setting. Leave the probe-wires plugged into the meter, but wrap them neatly around the meter, and either tie the probe ends into a loose knot, or secure them with a rubber band.

CHAPTER 3

Resistances in Parallel

RESISTOR TYPES

Methods for Obtaining Resistance

Resistance to the flow of electricity is greater when the wire or other conductor is very narrow, similar to the flow of water in a very narrow pipe. Electric current is really the motion of billions of electrons within the wire, and these all have negative charges, so they repel each other. The voltage is only sufficient to force a relatively small fraction of the *mutually repelling* electrons into a very narrow space at the same time, so the others must wait their turn, and this delay decreases the overall flow rate. The narrower the space, the slower the overall flow, at any given voltage. On the other hand, a higher voltage can force more electrons together into the narrow space, so that will tend to increase the flow.

The electrons hit various imperfections (including chemical impurities) in the wire and bounce off to the side to some degree, further slowing the overall forward motion. At higher temperature, the atoms of the conductive material are moving around randomly and are therefore not in their usual positions within the material, and this also slows the net forward motion to some degree. Also, the electrons tend to collide with each other, which also causes deviation from a straight path. Taken together, these effects are called "scattering," and that raises the electrical resistance that we measure with an ohmmeter. Examples of scattering effects are (1) nickel metal wire with 20% chromium mixed randomly into it, which acts as imperfections, and (2) heating the tungsten filament of a light bulb, and (3) making a wire narrow and thus causing more collisions within the narrowed space.

27

A fourth effect is that the electricity can be led through *one wire after another* (putting two wires "in series"), which doubles the resistance to flow, just like a very long pipe offers more resistance to the flow of water than a very short one does. In other words, the longer the conductive path, the greater the resistance.

All four of these effects are utilized in making various types of electrical resistors. The metal alloy 80Ni, 20Cr is commonly used to make "wire wound" resistors, usually wrapped around a porcelain ceramic insulator. These can be self-heated without damage, so they are useful where high power is needed. The heating effect is used in a special type of "thermistor" device, where temperature can be measured by noting the rise in resistance. The narrowness effect can be used by simply having a thin wire or a "thin film" of metal. More commonly, the narrowness effect is used by mixing graphite powder with a small amount of insulating clay — wherever the powder particles touch, the contact points are extremely narrow conductors, giving a fairly high resistance. High resistances can be obtained by making the devices quite long, but if extremely long paths are needed, the conductors are arranged in spirals or zig-zags ("meander patterns"), to save space.

The inexpensive resistors to be used in this course are mostly the "carbon composition" type, which are graphite plus clay, with a plastic insulating coating around the outside. These are quite similar to pencil "lead" (in this case pronounced "led" like the metal), where the 4H type is mostly clay and has a fairly high electrical resistance, while the 2S is mostly graphite and has a very low resistance. The insulating coating around the resistor is usually epoxy plastic, similar to the two-tube glue that can be purchased in a hardware store. Copper wires are embedded in the two ends, and the wires are coated with thin solder, giving them a gray color.

Standard RETMA Values

Ordinary resistors are only available in a few standard sizes, which have been agreed upon by the Radio-Electronics-Television Manufacturer's Association (RETMA). These sizes are various decades, such as 1, 10, 100, 1000, 10000, etc., multiplied by any number chosen from:

$$10, 15, 22, 33, 47, \text{ and } 68.$$

In other words, 10 ohm, 100 ohm, 1000 ohm, etc. values are available, and also 15, 150, 1500 ohm values, and 22, 220, 2200 values, and 33, 330, 3300 values, and so on. However, it is not easy to buy a 130 ohm or 140 ohm resistor, so the designers try to figure out how to make use of either a 100 ohm or 150 ohm value, or other RETMA multiples.

The series listed above consists of multiples of the sixth root of ten ($10^{1/6}$ or $\sqrt[6]{\ }$),which is 1.467. Multiplying that times itself six times yields a number close to 10. This allows us to divide a decade (that is, a group of resistances increasing by a factor of ten, like *100 to 1000*) into six small steps, spaced out evenly in a logarithmic manner (that is, like *100, 150, 220, 330, 470, 680, and 1000,* or in the next decade, *1000, 1500, 2200,* and so on).

Color Coding

Resistors are often labeled by little rings printed around the outsides, where each color represents a number. The ring **closest to one end** of the resistor is the **first** digit, such as the "1" of the value 1500 ohms. The next ring is the second digit, which is the "5" of the value 1500 ohms. So rings representing 1 and 5 would, taken together, stand for the number 15.

The third ring is not one of the digits — instead, it represents the number of "tens" multiplied against those first two digits above. An example would be, for the value 1500 ohms, the third ring would represent 2, meaning hundreds, so altogether there are 15 hundreds of ohms. (If it had been 3, it would mean thousands, or 15000 ohms.) Therefore, 1 — 5 — 2 would signify 15 multiplied against 100, or 1500 ohms total. Or, a 2 — 2 — 2 would mean 2200 ohms. Similarly, a 1 — 0 — 1 would mean 100 ohms

The numbers signified by each color are as follows:

Black		0	(Black is used in the middle ring.)
Brown	1	1	x 10 Br.-br.-br. = 110 Ω (not RETMA)
Red	2	2	x 100 Red-red-red = 2,200 Ω *
Orange	3	3	x 1,000 Oran.-oran.-oran. = 33,000 Ω
Yellow	4	4	x 10,000
Green	5	5	x 100,000
Blue	6	6	x 1,000,000
Purple	7	7	x 10,000,000
Gray	8	8	x 100,000,000
White	9	9	(White is not used for "tens.")

The fourth ring (not near an end of the resistor) represents the "tolerance" (accuracy). Gold = plus or minus 5% (usually written±5%), and silver = plus or minus 10%, and no colored ring there means plus or minus 20%.

---

* In Europe, 2k2 = 2,200, and 6M6 = 6.6 million. Also, 4R7 = 4.7 and 270R=270, where the R just stands for a decimal point.

TWO RESISTORS

Water Analog

If a water pump system is operated at the same pressure (height of water level visible in the glass tube) as was used in the previous chapter, but this time there are *two* faucets instead of one, then more water would be expected to flow. This is illustrated in Fig. 3.1. If each of the new faucets is open to the

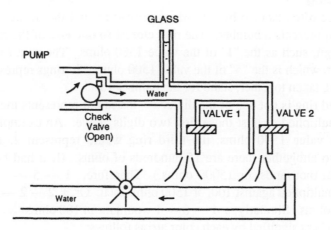

Fig. 3.1 A water pump analog of two resistors in parallel.

same degree as in the previous chapter, it is intuitively apparent to the reader that the total water flow rate will now be *double* that of the previous chapter (compare Fig. 2.1 on page 11). The two valves (which are resistances to flow) are said to be "in parallel." It is obvious that if the water flow is doubled, then the *total resistance* to flow must be *half* that of the single faucet.

Electrical Example

The reader can easily imagine an electric circuit where a battery is analogous to the water pump, and two resistors are analogous to the valves shown above. Of course, the flow of electric current would be double that of the case where there is only a single resistor, very similar to the behavior of the water analog. If the voltage is the same but the *current is doubled* in the two-resistor case, then Ohm's law tells us that the *total resistance* in the two-resistor case must be *half* that of the one-resistor case. This is quite intuitive and simple, when both resistors are equal.

If the two parallel resistors are *not* equal, however, we still can calculate the total resistance. An example is shown in Figure 3.2.

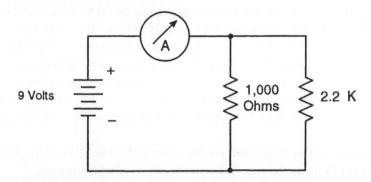

9 Volts

1,000 Ohms

2.2 K

Fig. 3.2 Two unequal resistors in parallel. (The circle is an ammeter.)

Just as the water flowing through two faucets adds up, so does the electric current through these two resistors. Adding the two currents, the total is:

$I_{total} = I_{large} + I_{small}$.   Then, using Ohm's law:

$$\frac{9\ volts}{R_{total}} = \frac{9\ volts}{1000\ \Omega} + \frac{9\ volts}{2200\ \Omega} = 9\ ma + 4ma. = 13ma = I_{total} \quad (3.1)$$

Now, we can use Ohm's law a different way, seeking the total resistance:

$$R_{total} = \frac{9\ volts}{I_{total}} = \frac{9\ volts}{0.013\ amp} = 692\ ohms. \quad (3.2)$$

OPTIONAL SECTION

It is interesting, and possibly surprising to the reader, that we do not need to know the voltage. We can divide equation 3.1 by the voltage, and then a *general* equation governing *resistances in parallel* will result:

$$\frac{1}{R_{total}} = \frac{1}{R_1} + \frac{1}{R_2} \quad (3.3)$$

Reciprocals of a third and fourth (or more) resistances could also be added to the ones in equation 3.3. The water analog shows intuitively why that is allowed: any number of faucets could be added, and if the pressure is held constant, they simply add to the total flow, and therefore they *decrease the total resistance* to flow.   Once all this is understood, equation 3.3 does not really have to be memorized. The reader only needs to know that the voltage could be cancelled out, giving the general form (equation 3.3, which is often the only form shown in other electronics books, and usually without explanation.)

The main things to remember are the ideas behind equation 3.1. Then it should be easy to figure out the total value of any parallel resistances.

EXPERIMENTS

Current

Looking at Fig. 3.2 on the previous page, a black clip lead can be attached to the negative battery terminal, using the careful method of Fig. 1.1 on page 2. Then "bottom" wires of the two resistors (brown-black-red and red-red-red as on page 29) can be crossed, and the other end of the black clip lead is attached to both of them at the same time, as in Fig. 3.3 below.

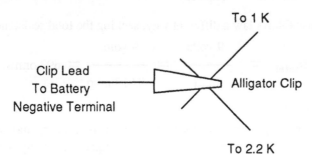

Fig. 3.3 Attaching a clip lead to *two* wires at the same time.

A red clip lead is then attached to the positive terminal of the 9V battery and also to the red metal "probe" of the multimeter. The rotary switch of the meter is set to "150mA DC" (or a similar range if a different kind of meter is used). The black probe of the meter is touched firmly to the other ("upper") end of the 1,000 ohm resistor. On the zero to 15 scale (which now really indicates zero to 150 ma), it will be observed that the current is about 9 ma. Of course, this is not an accurate measurement, since the scale is so compressed.

When the black probe of the meter is then touched to the upper lead of the 2.2K resistor, the current is only 4 ma, as predicted by equation 3.1. When the upper leads of both resistors are bent to be in contact with each other, and the black meter probe is touched to both of them simultaneously (possibly using an additional clip lead), the current is 13 ma, or something similar to that value.

Resistance

Since the Radio Shack inexpensive meter is not designed to measure low currents, it is somewhat more accurate to use it as an ohmmeter for this experiment. Disconnect the 9V battery, and just use the meter set to "RX 1KΩ." Calibrate it, measure the resistances of the 1K and 2.2K individually and then in parallel. Be sure that the multimeter is returned to the "off" setting.

THE AMMETER "SHUNT"

The inexpensive ammeter used in this laboratory course also can not handle high currents (>150 ma). If it is ever necessary to measure a large current, that can be done by putting a standardized "shunt" resistor of very low resistance in parallel with the ammeter. If the shunt has the same resistance as the ammeter (usually less than one ohm), then all the measurements visible on the original meter can be doubled. If it is 1/9 of the ammeter resistance, then the current going through this shunt will be 9 times the current in the original meter. This added to the original current would be ten times the original value, so that all the visible measurements could be multiplied by ten. Similar shunts can be arranged for other multiples. Expensive ammeters often have built-in, switch-selected shunts.

EQUIPMENT NOTES

Components For This Chapter	Radio Shack Catalog Number
Battery, nine volts	23-653
Clip leads, 14 inch, one set.	278-1156C
Multimeter (or "multitester")	22-218
Resistors, 1,000 and 2,200 ohms, 1/2 watt	271-306 or similar
Color code slide rule	271-1210

CHAPTER 4

Series Resistances, Part I: Bad Output Voltages

TWO RESISTORS

Just as the news media (TV, newspapers, and magazines) tend to concentrate on accidents and crimes, this book tends to concentrate on common industrial problems, in the hope that the reader will not become bored. Thus the word "Bad" in the title of this chapter. However, there is good news, too, because these problems can be solved. The solutions are often quite easy, provided there is some depth of understanding. Once again, several different analogies involving water behavior will be used as aids in explaining electrical behavior. In a few cases the result is not what one might ordinarily expect, without considerable analysis.

Water Analog

If a recirculating water system is operated at the same pump pressure as was used in the previous chapter, but this time there are *two* of those valves, one after another, then each one of the valves contributes some resistance to the water flow. Therefore *less* water would be expected to flow. A pump and valve system of this kind is illustrated in Fig. 4.1 on the next page. If each of the two valves is open to the same degree as it was in the previous chapter, it is

35

intuitively apparent to the reader that the total water flow rate will now be *half* that of the previous chapter. The two valves are said to be "*in series.*"

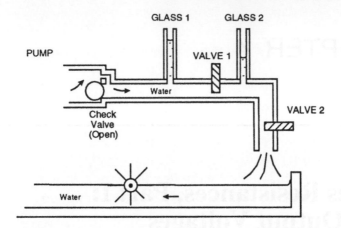

Fig. 4.1   A water pump analog of two resistors in series.

The diagram shows the two valves being *equally* open, but some important variations must be considered, as follows. (The pump pressure is constant.)

Example A: If Valve 1 remains only half closed, but Valve 2 is *closed almost all the way* (which is not really shown in the present diagram), then pressure would build up in Glass 2, and soon the pressures indicated in Glass 1 and Glass 2 would be almost the same. Note that there is *not much* long-term *flow* of water through Valve 1, because the water is almost stopped altogether by Valve 2. And there is *not much difference* in pressures. (Note again that we are assuming the original pressure supplied by the pump itself to be constant.)

Example B: If Valve 2 is now *opened* almost all the way, then there would be *more flow* of water *through Valve 1*, because there is less total resistance in the whole system. But if Valve 2 is opened so much, then the pressure indicated in Glass 2 would be *almost zero*, because the water could spill out easily. The **difference** between the Glass 1 and Glass 2 readings would be **great.** You could call that *difference* in the water level heights a "**PRESSURE DROP**" caused by the *resistance of Valve 1* (which is still closed half way).

There was *not much pressure drop* in Example A, because there was *not much flow* through Valve 1, but in Example B there was *more of both.* The lesson from these is: "THE MORE *FLOW,* THE MORE PRESSURE DROP."*

---

* **Advanced Note:** Valve 1 is analogous to the 100 ohm resistor in Fig. 4.2.. The opening of Valve 2 is analogous to attaching the dashed-line wire in that circuit diagram.

Example C: If Valve 1 is *opened* almost all the way, the flow will be great, but there will be hardly any difference between the readings in Glasses 1 and 2, because the water will just pour easily through Valve 1. The reason for *less difference* is that Valve 1 now has *less resistance*, compared to the previous example.

The lesson obtainable by comparing the Examples B and C: less resistance, less pressure drop. However, the opposite must also be true: "THE MORE *RESISTANCE*, THE MORE PRESSURE DROP."

Combining these capitalized words with those at the bottom of the previous page, wherever there is more *FLOW* and/or more *RESISTANCE*, there is more PRESSURE DROP. Now the reader should recall that in Ohm's law, IR = V, which says "when there is more *FLOW* and/or more *RESISTANCE*, there is more VOLTAGE." Therefore, the **voltage difference** of going one side of a *resistor* to another is analogous to "**pressure drop**" going from one side of a *valve* to another.

What appears to be possibly strange in the water analog, however, is that the pressure drop is not *causing* the flow, and in fact it looks like it is the other way around: the flow is causing the pressure drop. Now is a good time to recall the discussion at the bottom of page 18 — the mathematical description of the behavior is the same, regardless of which is thought of as *causing* the other. Either way, V = IR, whether voltage is the cause or the effect.

One could think of Example A as being an example of "**back pressure**," where the more that Valve 1 is closed, the more the pressure tends to escape up through Glass 1. In fact, that is a useful way to remember this little lesson. (Note also that in Example A, the more that Valve 2 is closed, even if only half way, the more pressure tends to escape up through Glass 2, so it works the same way for either valve.) "Back pressure" will turn out to be a useful way to think of "voltage drop," especially in the next chapter.

Electrical Example

The reader can easily imagine an electric circuit where a battery is analogous to the water pump, and *two* resistors are analogous to the valves in Fig. 4.1. Of course, the overall flow of electric current would be *half* that of the case where there is only a *single* resistor, just like the flow of water is halved when a second valve is added. If the current is only half, then Ohm's law tells us that the *total resistance* in the two-resistor case must be *double* that of the one-resistor case. This is quite intuitive and simple, if both resistances are *equal*.

Two resistors in series are shown in Fig. 4.2, but in this case the resistances are *not equal.* The circle with an arrow and the letter "A" inside it is an ammeter, just as it was in the previous chapter, on page 31.

The circle with a "V" in it is a voltmeter. Being an electrical component, it must be connected to *two* wires, in order to "complete the circuit" (Even in the case of the neon tester with only *one* wire plugged into the wall socket, on page 6, nothing happened until a person — possibly you — touched the other wire. That person was really "connected" to the ground, by "capacitance," which will be explained in a later chapter.) In Fig. 4.2, the negative wire at the bottom can be considered to be the "ground."

Each voltmeter in the figure measures a *difference* between two voltages. The bottom one, marked "3rd," is analogous to the "Glass 2" pressure indicator in the water system (Fig. 4.1), since it measures the difference between a wire and the "ground." In other words, looking at Fig. 4.1 again, Glass 2 is only connected to one thing, the pipe. However, it is not so different from the bottom voltmeter, because Glass 2 goes out into the air, and the air might be thought of as another wire, going down to the open trough at the bottom of the water system. The water could actually spill back down to the trough (analogous to the black "ground" wire in the electrical system), if the pressure got high enough to overcome gravity's downward pull on that water.

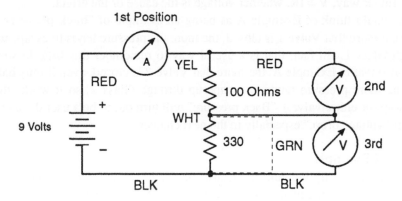

Fig. 4.2 An electric circuit with two resistors in series.

With the water system, we had to *imagine a difference* in the readings between Glass 1 and Glass 2. However, here in the electrical system of Fig. 4.2, the top meter ("2nd") *measures a difference* in voltages *automatically,* because it has two wires. As mentioned above, the bottom meter ("3rd") is also measuring a difference, but between the middle wire and the ground.

If we had three meters, we could set them up as in Fig. 4.2. However, we have only one for each lab group, so that single meter will first be set for *ampere* measurements, in the "1st Position" of the diagram, and no meter will be in the second or third positions at that particular time. Later, the same multimeter will be set for voltage measurement and put in position 2. Still later it will be put in the third position. (The importance of this "3rd position" will be appreciated when the reader learns about "Output 3" on page 170.)

EXPERIMENTS

Resistance
Find the 100 ohm (brown-black-brown) resistor. Note that with inexpensive resistors (the author will resist the temptation to use the word "cheap" or even worse adjectives!), "brown" often looks somewhat like purple or red, especially when viewed with fluorescent light or sunlight. Using the multimeter as an ohmmeter, and calibrating it for "zero ohms," measure the resistance, to make sure that the color code is being read properly. The measured resistance should be 100, plus or minus 10% if the fourth band is silver, which could be anywhere between 90 and 110 ohms. Then find the 330 ohm resistor (orange-orange-brown) and measure its resistance also. Set the multimeter to "OFF."

Connect the 100 Ω and 330 Ω resistors together with a white clip lead. Although it is not shown specifically in Fig. 4.2, use the multimeter as an ohmmeter to measure the resistance of this *pair of resistors in series*, either by attaching black and yellow clip leads to the multimeter and then to the bottom and top ends of the pair (black and yellow wires in Fig. 4.2, but no 9V battery is used), or else by simply touching the probes directly to the top and bottom resistor leads, without bothering with the actual black and yellow wires. The total resistance reading should be about 430 Ω. In other words,

$$R_{total} = R_1 + R_2 \qquad (4.1)$$

which the reader probably learned in high school physics (or maybe guessed, even without any books or teachers). Indeed, this is what one would expect as an electrical analog of the water pump system involving two valves in series, at least at very low (linear) flow rates. If more resistors were put in series, they would further increase the total resistance, also in an additive manner.

Current

Set the multimeter to "OFF." Looking at Fig. 4.2, the black clip lead should be attached to the battery first, and the other end of the wire is clipped onto the 330 ohm resistor. The top end of that resistor is attached to a white clip lead, and the other end of that lead is clipped onto the 100 ohm resistor. No voltmeters are attached at this time, but the multimeter is set for "150mA DC" (if it is the Radio Shack model), and its black probe is hooked up with a yellow clip lead to the top end of the 100 ohm resistor.

Whenever possible, it is a good idea to arrange the components on the lab bench in pretty much the same relative positions as shown in the diagram — this makes errors less likely. In the beginning, nothing is attached where the dashed line is shown.

The last connection to be made (and the first thing to be disconnected after the experiment is finished) is the red meter probe, touched firmly to the positive (smaller) battery terminal (or, alternatively, to a red clip lead attached to the battery). The meter should read about 22 ma. (Disconnect the battery **immediately**, to avoid overheating.) Using Ohm's law, it is apparent that 9.5V / 0.022 amperes = 430 ohms, which again is the sum of the resistances.

OPTIONAL

A green clip lead can now be attached **for a moment**[*] where the dashed line is on the diagram. The current increased, to 9.5V / 100 ohms = 95 ma, because one of the resistances has been bypassed. The green wire, which was used to bypass or "short circuit" the 330 ohm resistor, is sometimes referred to in electronics as a "jumper cable." It is now completely removed from the circuit. The meter is also removed and set to the "OFF" position.

COMMENT

These pages seem very close to being obvious, but the mechanics of assembling such circuits are not obvious to everybody, and in fact, many students make one or more mistakes in these assemblies. *Practice* will help. Also, a certain amount of confidence is useful in tackling important problems, and this confidence usually develops only with practice. It has been the author's experience that the lessons from the next experiments have been useful in solving "real-life" problems, but the lessons have been *far from* obvious to many workers, both in the lab and in the factory.

---

[*] Use page 17 to estimate wattage in the 100 ohms. See page 48 for wattage "rating."

Voltage

The next thing to do is set the meter to measure *volts* (dc, of course), in the 15 volt range. In order to put the voltmeter in the "2nd" position of Fig. 4.2, it is disconnected, and then its black probe is clipped to the junction between the two resistors, possibly by sharing an alligator clip on the white wire (Fig. 3.3, page 32). Its red probe is attached to the top end of the 100 ohm resistor, possibly by sharing an alligator on a red clip lead. Thus the voltmeter is said to be "connected across" that resistor.

When the red clip lead is attached from the battery to the 100 ohms (and the voltmeter), this situation will be analogous to Example A, at least in a rough approximation. The *difference* between Glass 1 and Glass 2 is like the voltmeter reading. The 100 Ω is like Valve 1, and the 330 Ω is like Valve 2. There is not much voltage, about 2.2 volts (compared to the battery's 9.5 volts), just like there was not much difference in pressures in Example A.

If possible, each reader should actually assemble this whole circuit without any assistance from anybody else. Generally speaking, if the result is not what was expected, the best strategy is to disconnect the red wire, and then look carefully at *everything*, starting with the black wire. Think of some hypotheses that might explain the unexpected result, such as "Maybe the battery has run down." Test each hypothesis, such as putting the voltmeter directly across the battery terminals (with the red clip lead still disconnected), or measuring the resistance with the ohmmeter setting, but *through* some of the clip leads that have been set up. Is the total resistance 430 ohms? If not, maybe there is a bad connection, or possibly a faulty clip lead, which quite often does happen.

It is important to realize that, if there were no voltmeters attached, the **current** flowing through the 100 is the **same** as the current through the 330, just like the water **flow** through Valve 1 is the **same** as through Valve 2. The voltmeter has a high resistance (about 15,000 ohms), so it passes very little current, and therefore the previous sentence is a good approximation of what is happening. If the current is essentially the same in both resistors, then we can use Ohm's law to tell us what the voltage is across each resistor:

$$(0.022 \text{ amp}) (\mathbf{100\ \Omega}) = 2.2 \text{ volts}, \quad \text{and} \quad (0.022 \text{ amp}) (\mathbf{330\ \Omega}) = 7.3 \text{ volts}.$$

What is happening is that the "**voltage drop**" across the 330 is being *subtracted* from the 9.5 battery voltage, leaving the 2.2V that we are measuring. That is why it is called "drop," because the voltage on the meter in the 2nd position is being decreased by that amount. It acts as if there is a small 7.3 volt battery where the 330 is, opposing the 9V battery. The top would have to be positive and the bottom negative, for the two voltages to oppose each other (subtract).

A simple way to think about this is that the 330 holds back some of the current, because it has a fairly high resistance. That is true, but a better way to picture what is going on is to realize that the 330, with some current going through it, has that voltage drop mentioned above (or a "back voltage," just like the "back pressure" in Example A). This is a better description of what is going on, because it is not just dependent on the amount of resistance in the 330, but also on the amount of current. It is the **combination** of resistance and current, according to Ohm's Law, that determines the **voltage drop**.

COMMENT

This discussion might seem like "beating a dead horse," since many readers already understand these things. However, the goal of the author is to make it likely that the reader will not forget this in the future, while working on a job. There is a strong tendency for students to remember principles for an exam but forget them immediately afterwards. Making mental comparisons to water pipes, and looking at things in different ways, have been found to enhance long-term memory.

The green clip lead is now attached **for only a few seconds**[*] across the 330 ohm resistor. This is shown by the dashed line in the diagram. The "2nd" voltage should rise to approximately 9.5, even though the 100 Ω is still attached. It is apparent from the diagram that the 330 Ω is no longer providing that "voltage drop," since it has been "short circuited," and therefore the voltmeter shows the full battery voltage. This is analogous to Example B.

Another way to look at this situation is to consider that the 100 and the voltmeter are now hooked up in parallel. Devices in parallel have the same voltage on them, just like *faucets* in parallel have the same *pressure* (page 30).

SUMMARY   Devices in **parallel** have the same **voltage** across them.

Devices in **series** have the same **current** going through them.

Quickly disconnect and then reconnect the 100 ohm resistor, while keeping the meter in the 2nd position and having the green wire attached across the 330. The voltmeter reading changes by a very small amount.[*] This will be explained a few pages later (bottom of p. 47). Now remove that green wire.

---

[*] Do not leave the 100 ohm resistor directly across the battery for a long time.

The multimeter is then moved to the third position in Fig. 4.2, and it will be left to the reader's ingenuity as to how to actually do this. The voltage reading will be, as expected, 7.3 V, with positive being at the top, as if it were another battery. Voltage "drop" can be an active thing, available as a voltage "source" to go somewhere else (like going to that meter in the 3rd position).

SUMMARY

> Any time current goes through a resistance, some of the driving voltage is decreased ("dropped"), following Ohm's law. However, that voltage is available to go somewhere else.

The Potentiometer

There is a mathematical formula for how much voltage was available in the "2nd position" of the meter. By the way, that was the *second meter* position, but it was attached across the *first resistor that the positive current hits,* as it goes around to the right and then downward toward "ground." Since it was the first resistor, we will call the 100 ohm resistor "R_1," and the 330 will be "R_2." The voltage across R_1 is called V_1. Then:

$$V_1 = V_{total} \frac{R_1}{R_1 + R_2} = 2.2 \text{ V} \qquad (4.2)$$

Similarly:

$$V_2 = V_{total} \frac{R_2}{R_1 + R_2} = 7.3 \text{ V} \qquad (4.3)$$

If there were more than two resistances, they could simply be added to the denominator (the bottom of the fraction).

Two resistors with a connection available between them (the white wire) can be called a "potentiometer." (Potential is another word for voltage — see the middle of page 14. The reason for the "meter" in the word is that this is sometimes used to measure voltage, but that will have to be explained in more detail later.) It can be used to divide a large voltage (9.5V in this case) into smaller parts (like 7.3V, etc.). Sometimes it is called a "voltage divider."

There is a component that allows the resistances to be varied quite easily, just by rotating a shaft or knob. This device is also called a potentiometer (or quite often it is just called a "pot"), even though it is really only a "variable resistor," and it is only rarely used as a "meter" for "potential."

Find the 5,000 ohm potentiometer, which is a round device with a metal shaft going into the middle, and three metal terminals attached to the edge. Hook it up as in Fig. 4.3. The red probe of the voltmeter goes to the middle terminal of the pot, and the two end terminals are hooked up "across" the battery. As usual, the red wire to the battery is attached last. Rotating the shaft all the way up* and all the way down makes the voltmeter read approximately 9.5 volts and then zero volts. This will show the reader what the limits are on the shaft rotation. The shaft is then set to nearly the middle of its range.

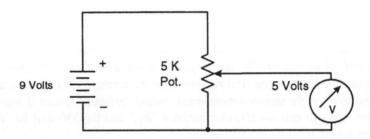

Fig. 4.3 A five thousand ohm potentiometer, in a test circuit.

It would be a good idea to draw a short line on the metal shaft of the 5K pot with a marking pen, so its rotated position is always immediately apparent.

If this pot is of the "linear taper" type, then according to equation 4.3, the voltmeter will read somewhere between 4.5 and 5 volts. However, some pots have an "audio taper," in which case the resistance versus rotation is logarithmic, and the 5 volt point will not be anywhere near the middle.

Most animal senses, including human hearing, respond to intensity in a logarithmic fashion. An "audio taper" (really a "log taper") pot used as a volume control in a radio or stereo system seems to the person using it as though its rotation is proportional to the perceived loudness, because both the pot and the ear have "log tapers." The logarithmic sensors in animals allow sensor responses to cover enormously wide ranges. The loudest sounds a human can stand without ear damage are about 10^{12} more energy-intense than the softest perceivable sounds. A person listening to this range would quite likely estimate that the loudest sound is "about ten or twelve times louder," not "a trillion times louder," but it actually would be the latter in terms of energy.

---

*   "*Most* of the way up" is analogous to Example A on page 36, but the voltmeter would be like looking at Glass 2 alone, without looking at Glass 1.

LESSONS

The following do not have to be done as actual experiments, since they are "lessons" that are logically apparent from the previous experiments. However, some readers might want to move a few resistors around to demonstrate the scientific principles firsthand.

A Power Source of Exactly 1.00 Volt

The author once had an experience where a colleague needed an exactly 1.00 volt power supply, in order to calibrate an apparatus for regulating temperature. He had no idea how to obtain this voltage, since a flashlight battery is about 1.5 volts. The author suggested the circuit of Fig. 4.3, using an accurate digital voltmeter, which is quite easily obtained. The pot was set to 1.00 V and periodically readjusted according to whatever the voltmeter indicated, and this worked very well. The apparatus, when calibrated this way, checked nicely with another instrument known to be of high accuracy.

Bad Batteries

In both Fig. 4.2 and Fig. 4.3, let us imagine that the *upper* half of the "potentiometer" is the resistance of the *internal* guts of the battery, and the *lower* half of the pot is the thing being *operated* by the battery, possibly an electric motor, or a heater, or a light bulb. In electronics, the thing being operated is often called the "load." Now suppose the "internal resistance" of the battery is very low, say 1Ω, and also the *load* resistance is quite *high*, say 1K. Then the voltage across the load is almost the entire battery voltage. An actual battery is supposed to have a fairly low resistance inside, sometimes less than one ohm. Hopefully, its effect will be small, as in this 1Ω and 1K example, where there is not going to be any problem, because the desirable 9 volts or so is available across the load.

Now consider a case where the *load* resistance is *unusually low*, like the resistance of a high power heater would be, possibly 1Ω or so. Then equation 4.3 tells us that the voltage "dissipated" in the load (the voltage being used in the heater) drops to only about half the battery voltage. The other half of the power would be lost as undesirable heat inside the battery, possibly damaging it by overheating. We want almost all of the battery's available energy to go into the load, not into the battery's own resistance.

This case of a low load resistance is analogous to Example B on page 36. Also, it is similar to the attachment of the green wire on page 38. The direct cause of the problem of internal power loss is that the current is high, and

Ohm's law (V = IR, having an impact once again) says that the voltage drop is fairly high if the current is high, because the R is never quite zero.

One can look at the problem in either of two ways, both correct. The undesirable voltage drop across the *internal* resistance can be thought of as too *high* because there is *too much current* ($V_1 = I\ R_1$, like one of the equations on page 41), or the desirable voltage drop across the *load* can be seen as being too *low* because the load *resistance is too low* ($V_2 = I\ R_2$).

The Rheostat

Figure 4.4 shows no connection to the bottom of the variable resistor. The "input resistance" of a voltmeter is usually quite high — in fact, the higher, the better.* Therefore the current in the diagram is low, and the voltage drop within the variable resistor is also low. (The internal resistance of the battery is not being considered here, because it is much lower than those shown in the diagram, and also the current is low.) Therefore the voltmeter will show almost the entire original battery voltage. This will be true whether the variable resistor is set for low resistance (upward in this drawing) or high resistance (downward). When the variable resistor is used in this manner, without any connection to the bottom terminal, it is not called a pot, but instead we refer to it as a "rheostat." It is often used to control current, but it does not have much *direct* effect on voltage.

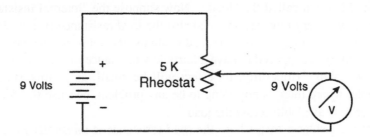

Fig. 4.4 A rheostat, which does not control voltage.

What is the reason why disconnecting the bottom terminal stops the rheostat from strongly controlling voltage, at least as it is drawn here? One way to look at the answer is that, with a very high resistance voltmeter, there is very little current going through the top half of the variable resistor, so there is

---

* The inexpensive voltmeter used in this course actually has an undesirably low R.

very little voltage drop. On the other hand, if we did later re-attach the bottom terminal to the "ground," going back to Fig. 4.3, then there would be significant current flowing through the resulting potentiometer *at all times*, regardless of the setting of the middle contact, high or low. If the middle terminal (the arrow) were set down low, then this current would have to pass through a fairly high resistance, and there would be enough voltage drop to lower the output voltage. (By the way, a voltage drop is sometimes called an "IR drop.") In other words, when we have current going through a resistor as in Fig. 4.3, then we will get a voltage drop, and this can be used to control an output voltage. Without that significant current, as in Fig. 4.4, then there is practically no effect on the output voltage.

Another symbol for a rheostat, equivalent to the pot with only two connections in Fig 4.4, is shown below in Fig. 4.5. The open circles are "terminals," which can be connected by any practical means, such as plugs, clip leads, etc.

Fig. 4.5 An alternative symbol for just the rheostat alone.

High "Upper" Resistance in the Potentiometer

Going back to Fig. 4.3, another common problem is that the voltage drop across the *internal resistance* of a battery is sometimes too *high*. With an automobile battery in very cold weather, the internal resistance is nonlinear, like that of a light bulb. At the high current levels necessary to operate the electric "starter" motor of a car, the nonlinear internal resistance of the battery can become quite high, and then there is not enough output voltage (voltage across the load) to get the engine started. This is different from the light bulb case, in that the nonlinearity is not caused by high temperature — it is caused by low temperature. But it is still nonlinear with respect to voltage versus current. At any rate, the internal resistance of any kind of electric power source should be kept as low as possible, so that currents can be supplied without significant decrease in the output voltage.

Back on page 42, at the very bottom, the effect of a moderate amount of internal resistance was noted. Quickly disconnecting and then reconnecting a 100 ohm load made the voltage drop of the battery's internal resistance become visible as a slight change in output voltage.

EQUIPMENT NOTES

Components For This Chapter	Radio Shack Catalog Number
Battery, nine volts	23-653
Clip leads, 14 inch, one set.	278-1156C
Resistor, 100 ohm, 10 or 1 watt	271-135 or 271-152A
Resistor, 330 ohm, 1/2 or 1/4 watt	271-306 or 271-308
Potentiometer, 5K	271-1714 or 271-1720
Multimeter (or "multitester")	22-218

Note that if "leads" (the wires coming out of a device such as a resistor) have to be bent, they should come out straight for a short distance and then only be bent at the place a few millimeters away from the device itself. On the other hand, if they are bent right at the edge of the device, then the insulation around the device is likely to become cracked.

If the experimenter has not already done this while reading the middle of page 44, it would be a good idea to draw a small line on the long metal rod of the 5K pot with a marking pen, so its rotated position is always immediately apparent.

CHAPTER 5

Series Resistances, Part II:
Bad Measurements

EXPERIMENTS

In the previous chapter, the experiments of Figs. 4.3 and 4.4 worked just as they should. Now let us try a similar setup, the one in Fig. 5.1, with higher resistance values. At first, just use the voltmeter to the right, and do not hook up anything where the left-hand voltmeter and switch are shown.

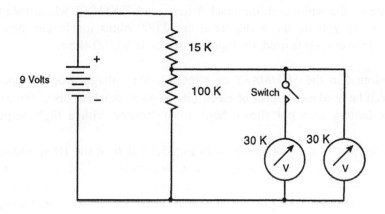

Fig. 5.1 A voltage source with high output resistance, measured by a meter with a low input resistance.

The voltage measured should be around 9.5V(100K / 115K) = 8.3V, according to equation 4.3 on page 43. However, your actual measurement will be significantly lower — more like 6 volts.

The experiments in the previous chapter were purposely designed to utilize very low values of resistance, in fact so low that the reader had to be cautioned not to leave the battery attached for a long enough time to overheat the resistors. In this chapter the resistors are relatively high values, so excess current is not a problem. A different problem appears in this case: low voltage. To illustrate it, attach another voltmeter, borrowed from a neighboring experimenter, or else use a resistor of 22K or 56K ohms. (Note that 56K is not a standard RETMA value, as listed on page 28, but it is often supplied in "assortments" of resistors anyhow.*) Alternately connecting and disconnecting this extra meter, using a clip lead instead of a real "switch," will show that the second meter pulls down the voltage even further, to about 4 volts. This is a hint that the first meter probably decreased the indicated voltage also.

OPTIONAL

If another voltmeter can be borrowed, set it as an *ohmmeter,* in order to measure the resistance of yours set as a *voltmeter*, red probe to black, and black to red. The voltmeter will be observed to be about 30K, as noted in the figure. On a Radio Shack meter (Catalog number 22-218), at the lower left corner of the white dial, the black letters read "2KΩ/V," which means that for every volt on the scale, there are 2,000 ohms inside the meter. Since the 15 volt scale is used, the input resistance is 30,000 ohms.

As mentioned in the *SUMMARY* on page 43, the voltage drop in a potentiometer can be used as a *source* of electricity to go somewhere else. We can say that the battery plus pot shown here are a "source with a high output resistance."

The bottom part of the potentiometer in Fig. 5.1 has *both* the 100K and the 30K in parallel (using just one meter), so it is now only about 25K (see eqn. 3.3 on page 31) instead of 100K. Therefore the "output" voltage of the battery plus pot is only about 9.5V(25K/[15K + 25K]) = 6V. This is why the voltage reading observed in the experiment was so low.

---

* The reader is encouraged to use the Index, to find items like "RETMA," or "watt," etc.

The meter with a low input resistance has influenced the voltage being measured, which of course is highly undesirable. The lesson from this is: "the higher the better," at least for the input resistance of a voltmeter, so as not to influence the voltage being studied. A more general rule, which is not always true but often is a good idea to follow, reads:

"Use a *low output* resistance to drive a *high input* resistance."
(The "input" of the voltmeter is the "load" for the potentiometer.)

This way, the thing being operated (or "driven"), like the voltmeter in this case, is less likely to have any unpredictable effects on the thing which is supplying the driving force. (There are several exceptions to this rather loose guideline. One exception would be where very high currents are needed, and therefore the input *also* has to have a low resistance. Another one will be discussed on page 79, and still another will be discussed toward the end of this book, on page 213.)

The author has witnessed several incidents in which low input R caused trouble, when the person doing the measurements did not understand these effects. High quality meters usually have input resistances which are almost infinite. However, they tend to be expensive, and they can be damaged easily by static electricity, if they are not designed well.

To stand back for a while and put all this in perspective, we should recall that, at the bottoms of pages 45 and 47, *very* low load resistances pulled down the output voltages, even though the output resistances were *moderately* low, but not low enough for heavy duty use. In this chapter, the load resistances are higher (the inputs of inexpensive voltmeters), but they are still pulling down the voltages, because the output resistances are also high. Once again, the "general rule" quoted above, on this page, is often a good idea to follow.

THE WHEATSTONE BRIDGE

One of the best ways to measure an output voltage, where there is a *very high* output resistance, is to "match" that output voltage with another, purposely-created voltage, so the two voltages are equal and facing each other. Then essentially zero current will flow into the "meter," and there is negligible IR drop, even with high resistances all around. The circuit used to carry out this idea is called the Wheatstone bridge, named after the early scientist who suggested it. Sometimes it is just called a "bridge circuit." It is shown on the next page, in Fig. 5.2. (Electronic engineers tend to say *"impedance"* instead of "resistance." That term has special meaning with alternating current, but with dc as used here, the two words are interchangeable. Impedance with ac will be discussed later. Another term is "source impedance," meaning the same as "output resistance.")

The parts inside the dashed line are often available as a pre-assembled unit, with a carefully calibrated variable resistor that is marked in "volts" (from 1 to 4 in this diagram). The battery is a special type that puts out a predictable voltage, and all the parts are designed so they do not change much with temperature. The microammeter, sometimes called a "galvanometer," is not calibrated, but it is very sensitive. This whole assembly inside the dashed line is called a "potentiometer," and of course that is where the other types of (uncalibrated) variable resistors got their names.

To measure the unknown voltage generated at the left, a person operating the right-hand "potentiometer" adjusts the calibrated variable resistor until the sensitive ammeter does not show any visible reading. Then the voltage reading can be seen from the scale that indicates how high or low the resistor has been set. (In this diagram, it looks like about 2.5 volts.)

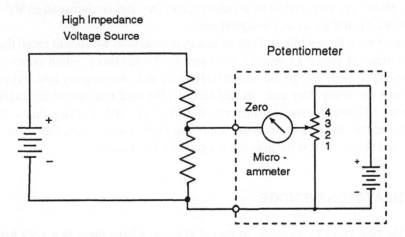

Fig. 5.2 The Wheatstone bridge for accurately measuring a voltage.

Several kinds of electronic devices, such as thermocouples for measuring temperature, have high internal resistances. Any appreciable current drawn from them will cause severe voltage drops, but if no current is drawn, the voltage outputs are very predictable. The "potentiometer" as shown above, arranged in the Wheatstone bridge circuit, is a good way to use these devices.

It should be mentioned that a thermcouple is similar to the "high impedance voltage source" in the figure on the previous page, except that there is no "bottom" half of the potentiometer. There is just simply an internal resistance, sort of like the rheostat on page 46. It is very nonlinear, and the voltage drop increases sharply when current flows.

Nowadays, hand operated, calibrated pots are mostly used only for specialized research. In factories, and for the more routine research measurements, an automatic variation is commonly used. The calibrated pot is set by an electric motor, and this motor stops automatically when the current becomes zero. A pen with red or other colored ink is attached to the pot, so the higher the voltage reading, the farther the pen gets positioned *across* a long strip of paper. This paper strip is driven *lengthwise*, by a clock motor. Thus the pen leaves a trail of voltage input versus time. The whole assembly is called a "strip chart recorder."

If the motor of a strip chart recorder moves too fast, it might "overshoot" the point of zero current and then swing back again, "hunting" for the best position. Therefore the motor is purposely slowed down, in order to "damp" its action. If, on the other hand, the damping is excessive, then the recorder will not be able to follow changing input voltages without error.

Any current reading other than zero in the ammeter is called an "error signal." The sensitivity of the ammeter can be adjusted, because too much sensitivity is likely to respond to random noise.

These types of adjustments are also important in other "automatic control" units, such as thermostats and automatic pressure regulators. They are also used in home oil heating and air conditioning systems. The principles will be explained further in later chapters.

EQUIPMENT NOTES

Components For This Chapter	Radio Shack Catalog Number
Battery, nine volts	23-653
Clip leads, 14 inch, one set.	278-1156C
Resistors, 15K, 56K, & 100K, 1/2 or 1/4 watt	271-306 or 271-308
Potentiometer, 5K	271-1714 or 271-1720
Multimeter (or "multitester")	22-218

CHAPTER 6

Series Resistances, Part III:
Bad Grounds

TRUE GROUND

Many (probably most) electrical circuits have one side of the power supply attached to the "ground."

(1.) **One reason** for doing this is that the device often includes some kind of a large conductor such as a metal cabinet, and the designer might as well make use of it to be one of the two necessary conductors, just as a convenience. If *direct current* is being used, the "ground" is often made the negative side, although it is not always done this way. If long distance *alternating current* systems are being used, such as the electric power in your house, the surface of the earth itself can be utilized for one side of the circuit. This will be a low resistance path, on account of its being extremely wide and thick.

(2.) A **second reason** for using the ground is for safety. The interior wiring will be carrying various voltages, insulated from the cabinet, as shown by Fig. 6.1 on the following page. (The reader should note the standard symbols for the grounds and for the source of alternating current, such as a 120 volt ac wall socket.) Imagine that the dashed-line "green" wire and the "leakage" are not present, and a lot of current is flowing through the white wire. Then the voltage drop across the white wire's resistance might make a considerable potential appear on the cabinet.

A person operating the equipment might accidentally touch the cabinet, and at the same time be touching a metal sink or light fixture, etc., which could be a good ground. If that person makes up a *better* ground than the white wire is, he or she might get a shock. (If the cabinet is plastic, a face plate or knob might be metal.)

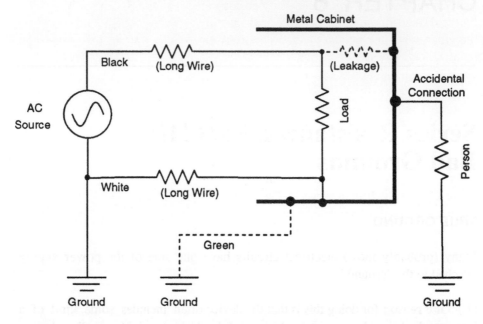

Fig. 6.1 A shocking situation. (Note symbols for ac source and ground.)

This situation would be even worse if other users of electricity (not shown in the figure) are pumping current through that same long white wire, back to the power source. Then extra voltage can appear on the cabinet, even if not much current is being passed through the local "load." This actually does happen occasionally, and it can be worse than if the cabinet was not "grounded" (that is, not connected internally to the white wire).

(2-A.) To **prevent** this type of bad grounding from having some serious consequences, a third wire is usually supplied in modern electric power systems, with green insulation and a round (instead of flat) female connector in the socket. This is not meant to carry any current unless something goes wrong with the insulation, and it is available to be a good ground connection to take any stray voltage directly to the same ground that people are standing on. The green wire does not go all the way back to the power source, but instead it is connected to a nearby ground such as a water pipe. Its voltage drop is practically zero, if its resistance is nearly zero.

An additional dangerous situation can exist if the wiring inside the cabinet has faulty insulation, and some electricity accidentally goes to the cabinet, as shown by the dashed-line resistor marked "leakage" in the figure. Once again, without the green wire, the person touching the cabinet is dependent on the white wire being a good ground. If the white wire is very long, and/or there are many other users, a dangerous shock can result.

(2-B.) Another **preventive** measure, being used in more and more modern houses, is a "ground fault interrupter." This is an automatic switch inside the wall socket which turns off the electricity if more current flows through the black wire than through the white wire. That excess current might be following some dangerous route like through a person and down directly into the ground, without going back along the proper "neutral" wire path. A portable ground fault interrupter will probably be used by the instructor when 120 volts ac is needed, later in this course.

(2-C.) Still another **preventive** measure is legal in the United States instead of using the green extra ground wire, and that is "double insulation." If the entire machine, such as a hand held drill, or a kitchen mixer, is coated with two independent layers of plastic insulation, that is considered safe enough to be used with only the black and white wires.

(3.) A **third reason** for having one side be grounded, including the exterior metal cabinet, is that this tends to prevent RFI (see Index or page 9). Radio waves coming into the device (TV set, computer, motor controller, etc.) from outside can interfere with the operation of that device. Nowadays, with satellites, cellular phones, TV, and radio, literally everywhere in the world, we are bathed in radio frequency ("RF") electromagnetic waves. If RF can create a voltage in an antenna, it might also do it inside your computer, possibly changing some important data. However, if the metal cabinet is grounded, this voltage goes harmlessly into the earth. If the cabinet is not made of conducting metal, it will probably be coated on its inner surface with a thin film of metal, or else a metal mesh, and of course these are then grounded, often via the green wire. This use of specially conductive layers is called "shielding."

These stray radio waves are sometimes called "electromagnetic fields," or "EMFs." (The old meaning of EMF, still used occasionally, is described on page 14.) If radio waves somehow got through the grounded-cabinet shielding of your computer, the problem would be called electromagnetic interference (EMI).

STANDARD CONNECTIONS

The wall socket used for "120 VAC" in the USA (but different in many other countries), is shown in Fig. 6.2, as viewed by a person standing in the room, looking at the wall. (The last statement is important, because some types of socket diagrams are drawn as viewed from behind.) The neutral wire (*usually* grounded) in houses and offices has white insulation, and it is the large rectangu-

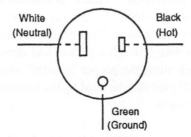

White
(Neutral)

Black
(Hot)

Green
(Ground)

Fig. 6.2 Standard wall socket ("power outlet" or "mains") for 120 volt ac.

lar connection in the wall socket. The "hot" wire is the smaller rectangular connection, and the insulation color is standardized as black. Electricians usually refer to the white wire as "neutral," rather than ground, because it is *not necessarily* a very good ground, even though it *is necessarily* a good path back to the power generator. (In England, "grounding" is often called "earthing.")

> **NOTE:** *Black is not* the color code of "ground" in **electrician's** ac wiring, although *black is* often the unofficial color code for the "ground" in the **electronic** engineer's dc wiring, *if* the ground happens to be *negative*. Black is negative for guitar and other musical ("PA ") loudspeaker wires.

> Unfortunately, *some* **thermocouples** use red as the color code of negative.

> Also, *some* local **telephone** systems use red* for negative (the "ring") and green* for positive (the "tip"), with neither being grounded. Some other phones are not polarity-sensitive and use the opposite colors.

---

* Because of occasional color blindness, blue with white dots is beginning to be used instead of red for modern *telephone* wiring, and white with blue dots instead of green.

THE CHASSIS

Inside the cabinet, there is usually a flat plane of aluminum metal on which many "components" (such as resistors, lamp bulbs, etc.) are mounted. It is called the "chassis" (pronounced "chassee"). This is usually "grounded," and quotes are used here because it is not always a "good" (low resistance) ground, but the designer assumes that it is a good ground. One end of the component is sometimes connected directly to this metal plane, for convenience. In recently made circuits, the chassis is more often a thin copper film attached to a plastic "printed wiring board," instead of being thicker, self-supporting aluminum. The chassis is identified in Fig. 6.3 as being separate from the true ground, because there sometimes is some resistance between the two.

There is usually a thick wire coming from the other, "hot," end of the power source (battery, or ac socket, or other power supply), and this main wire is called the "bus." (In England it is often called the "rail.") In Fig. 6.3, it is shown as being a positive-charged dc bus, and various components (the 330 and the 2.2K) get their electric power from this wire.

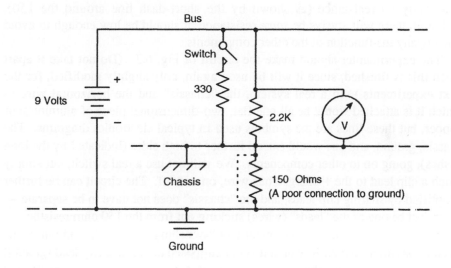

Fig. 6.3 One component interfering with another's operation ("SSN").

EXPERIMENTS

Simultaneous Switching Noise ("SSN")

In Fig. 6.3 on the previous page, the voltmeter and the 2.2K resistor represent much-simplified parts of a complex electronic system such as a computer. The "switch" and the 330 represent another part of the system, not connected to the 2.2K, except that they share the same power source and ground. When the switch turns on, it should *not* affect the voltmeter.

In a complex system, there are bound to be times when more than one of the switching components (usually transistors, rather than mechanical switches) turn on at the same time. If so, they will all be drawing power from the bus and chassis simultaneously, if they are set up in parallel, which they often are. If one (the 330) causes an additional voltage drop across an undesirable resistance that is in the ground connection (the 150), then this action might cause the voltage available for the other (the 2.2K) to be temporarily too low for proper operation. This is called "simultaneous switching noise" (or "SSN") in computer technology, or "common impedance coupling" ("CIC") in music recording and playback.

The solution to this problem is to make the connection to the true ground have practically no resistance (as shown by the short-dash line around the 150). Although there will always be *some* resistance, it should be low enough to avoid causing any malfunction of the other components.

The experimenter should make the circuit of Fig. 6.3. (Do not take it apart when this is finished, since it will be used again, only slightly modified, for the next experiments.) In a real system, the "chassis" and the horizontal wire to which it is attached would be all one flat, two-dimensional piece of aluminum or copper, but these lines are the symbols used in typical electronics diagrams. The chassis, ground, and bus would extend further to both sides (indicated by the long dashes), going on to other components. We will not use a real switch, but simply touch a clip lead to the + charged bus wire, on and off. The circuit can be further simplified by sharing clip leads, and the "chassis" does not have to be separate — it can just be one of the "leads" (wires) sticking out from the 150 ohm resistor.

It can be seen from the experiment that "switching" the 330 in and out of the circuit *does* affect the 2.2K, which it is *not* supposed to do. If a clip lead (green if available) is put where the short-dash line is shown, then there is hardly any effect, which is the desired result of having a "good ground."

Ground Loops

Looking at Fig. 6.4 below, imagine that the 330 is a powerful electric heater. There is a high impedance thermocouple (not shown) near the heater, and its "signal" goes to an automatic potentiometer (see Fig. 5.2 on page 52). This might be part of a strip chart recorder (middle of page 53), to keep a record of the temperature versus time. This can also be used to change the amount of power going into the heater, by means of other circuitry not shown here (but described later in this course). If the temperature gets too high, the switch going to the 330 is then automatically turned off ("opened") for a short time, and "closed" if it gets too cold. The whole thermocouple and pot system is called a "controller," and it is symbolized by the short-dash rectangle in the right half of Fig. 6.4. It can keep the temperature within a narrow range, like the "thermostat" of a heating system or air conditioner.

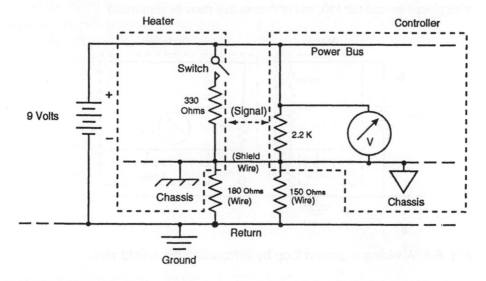

Fig. 6.4 A ground loop. (Some authors use other alternative symbols, not shown here, such as a black triangle to indicate the true ground.)

Presumably, each chassis is fairly well grounded, and the heater will not affect the controller, except via the signal from the thermocouple. However, some of the power going through the heater can quickly go to the controller through the "shield," before the heat gets a chance to build up and operate the thermocouple. This would be a false kind of "signal" which causes the system to malfunction.

The way such systems are usually hooked up, the signal from the thermo-couple is carried through a wire, surrounded by plastic or rubber insulation, and there is wire mesh *shielding* all around that insulation, to prevent RFI and EMI (see bottom of page 57). At *both* ends of the cable, this mesh is sometimes connected to each chassis, to "assure grounding." But that can be "too much of a good thing." In Fig. 6.4, it can be seen that there is a *complete circle,* going through the true ground itself ("return"), the 180 and 150 resistors,† and the shielding around the thermocouple wire. This is called a "ground loop," and it is something that good electronic designers try to avoid.

The experimenter should build the circuit of Fig. 6.4. At this point, things are getting a bit complicated.* The result of closing and opening the make-believe switch should be similar to the result of the previous experiment, except that there is slightly less effect. If this is not what happens, possibly go back to the previous circuit, try it again, and then re-attach the 180. (In Fig. 6.3, there was a dashed wire placed around the 150, and of course this must be removed.)

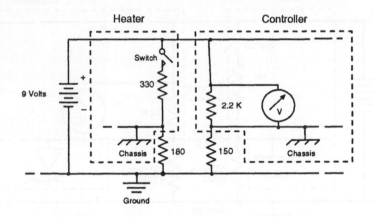

Fig. 6.5 Avoiding a ground loop by eliminating the shield wire.

The disadvantage of the ground loop, in real factory or laboratory equipment, is that the shielding (or other connection) is carrying stray voltage over to the controller. Therefore, two solutions to the problem are evident.

One is to make sure the shield (or any other inter-chassis connection) is disconnected, as in Fig. 6.5. The result is nearly obvious, but the experimenter

---

† In real systems, these might be long wires, and therefore poor quality ground connections.

* Electronics professionals sometimes call assemblies of wires like this "haywire circuits." A newcomer might be tempted to call this one a "rats nest."

should still go through the steps of doing this, partly just for practice in following circuit diagrams accurately. There should be no effect when the 330 is taken in and out of the circuit via the make-believe switch, except for an extremely small variation due to the internal resistance of the battery. The lesson from this is that separate things such as a heater and a controller can both be connected to "grounds," even imperfect ones (the 180 and 150), but they should *not also share another ground* connection such as the outer shielding conductor of a cable. If the shielding must be grounded somehow, to prevent RFI, which is usually the case, then *only one end* may be grounded (either end, but not both ends, at least not to a chassis). This is not shown in the diagrams, but it is often a useful "fix."

Neither the shield nor the signal wire are shown in Fig. 6.5, for simplicity, but they both could be used, as long as they don't connect one chassis to the other chassis. The *signal* connection is not a problem regarding ground loops, because it is only attached to the middle of a high resistance potentiometer, not to either chassis.

Solution number **two** for the problem is shown in Fig. 6.6. In case the shield connection can not be avoided in any practical way, or in case there is some other inter-chassis connection that can not be removed, all grounds can be made at a *single* point. Also, a great effort should be made to have that ground connection be a "good" one, with very low resistance. The experimenter should follow this diagram also, and observe that connecting the 330 has practically no effect.

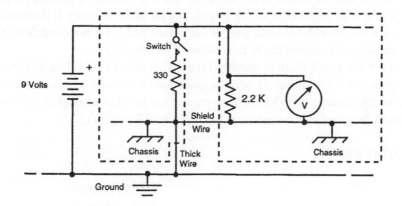

Fig. 6.6 Avoiding a ground loop by using a single, good connection.

Home "stereo" audio systems, as well as 5.1 channel DVD "home theaters," sometimes develop problems from ground loops. Usually the symptoms are low frequency "oscillations," where an on-and-off voltage goes to the loudspeakers with possibly destructive effect. The same two solutions to the problem are possible: (1) cutting the shield connection at *one* end of the inter-chassis cable, or (2) making sure that all ac plugs go to a *single* well-grounded cluster of sockets ("power strip" or "outlet strip"). This will be discussed further on page 82.

GUARDS

Whenever *very low* voltages or currents need to be measured, various electro-magnetic fields from ac power lines or from radio waves tend to cause random "noise" to appear on the "probe" wires of the measuring instrument. Instead of ac, sometimes this "noise" is dc, from electrochemical differences between conductors acting like tiny batteries, or from thermoelectric differences acting like tiny thermocouples. Such noise sources might have more magnitude than the low voltages being measured.

To prevent noise from randomly affecting the measurements, a "guard" conductor is sometimes used. It is similar to the "double insulation" mentioned in section (2-C.) on page 57, but here there is also a double conductor. One conductor, the "guard," is in between the two layers of insulation. It is not grounded, since the ground might have some stray voltages on it. Instead, in some cases it is not connected to anything, so that any voltage appearing on part of it tends to spread around the whole guard, and its local intensity is decreased. In other cases it is attached to some part of the circuit that is far removed from the input. Sometimes it is connected to the output of an amplifier.

Outside of the guard there is another layer of insulation, and then a "shield" layer of metal that is grounded. In some applications the whole structure is three-dimensional (concentric cylinders of wire mesh), but in other examples it is two-dimensional (circles of metal stuck onto a flat plane of insulator).

EQUIPMENT NOTES

Components For **This Chapter**	**Radio Shack** **Catalog Number**
Battery, nine volts	23-653
Clip leads, 14 inch, one set.	278-1156C
Resistors, 150, 180, 330, 2.2K	271-306 or 271-308
Multimeter (or "multitester")	22-218

CHAPTER 7

Soldering

STRIPPING INSULATION FROM WIRE

Obtain a power cord with an attached plug, either two- or three-prong. If it has already been soldered onto a transformer, cut it free by using the wire stripping tool as a scissors, close to its rotatable hinge.

On both the wires of the power cord, strip off the insulation, to about 3/4 inch (or about two centimeters) from the ends. This is done in the following manner. Use the wire stripper as if it was a scissors, but place the wire farther from the rotatable hinge, so the wire falls down into the pair of V-grooves. Try to cut into the plastic (or rubber) insulation, but not into the copper metal wire itself. (More resistance to the cutting action will be felt when you begin to hit the metal.) Release your grip on the tool, and the spring should open the tool's blades (jaws) part of the way — if not, open the blades yourself, part-way. Keep the tool at the same place on the wire, lengthwise. Rotate the wire about 45 degrees, and cut down into the insulation again, trying not to cut the metal. Release, rotate, and cut again, several times, so the cut you have made goes all around the insulation, separating the last 3/4 inch cylinder of insulation from the rest of it.

While keeping the blades closed part-way down into the round cut you have just made (but not pressing onto the metal very much), strongly push the tool toward the end of the wire, so the little cylinder of insulation slides entirely off the wire. The metal will now be exposed, and it is probably ready for soldering. In case, however, the metal appears black or otherwise dirty, scrape it clean with the sharp blades of the tool. (The loss of a few thin copper filaments is OK.)

MAKING A GOOD SOLDER JOINT

Find the heavily insulated black "primary" wires of the transformer (see pages 3 and 8). Make sure the insulation is stripped off the ends of the wires. Hold one power cord and one transformer primary wire, end-facing-end, and with the bare metal wires overlapping and next to each other. With your fingers, twist the bare wires around each other, but do not include any insulated parts. Put some cardboard or heavy paper on the table top, to protect it from damage. Arrange the twisted pair of wires so it is suspended an inch or so above the cardboard or paper, and it can later be touched from underneath by the hot soldering iron.

Place the soldering iron on the cardboard or paper, with the iron's tip resting on the L-shaped metal stand that was supplied with the iron. Plug the iron's power cord into a 120 volt ac wall socket or a thick extension cord. After about five minutes, touch the end of a strip of solder to the hot iron's tip, in order to be sure that it has become hot enough to melt solder quickly.

If a lot of black material or other dirty coating is on the iron, scrape that off with the sharp blade of the wire stripper, replacing the dirt with shiny, clean-looking melted solder. Add solder until a drop of it is sticking to the iron. Excessive solder might drip off the end, so make sure it only lands on the cardboard or paper. Melted plastic ("flux") might also drop, as well as evaporating upwards in the form of white smoke, but this is quite normal.

Note that the phrase "hot tip" is meant to be *not* the very end which is a sharp point. Instead, the soldering action is done with the fairly flat cone that is next to the very end, and also the cylindrical part next to that, if it is hot enough.

Place a stick of solder against the side of the twisted pair of wires, as shown in Fig. 7.1. Then, with a drop of melted solder sitting on top of the cone or cylinder of the iron, lift that drop upwards from under the twisted pair of wires, touching the wires with melted solder. The solder that is already on the iron will heat the larger stick of solder that is being held against the wires, and the "flux" (a special type of organic material that is inside the solder stick) will then quickly melt. The flux will "wet" the wires and dissolve off any thin layers of copper oxide that are on the wires. It will also prevent new oxide from forming when the wires get hot, so the flux is necessary for good contact of solder-to-copper, without any non-conducting oxide in between.

The next step is that the solder itself in the large stick will melt. Since the flux has already (hopefully!) cleaned the copper, the newly melted solder spreads across the surfaces of the wires and in between them. When it cools, after the iron is removed, the solder makes an excellent joint between the two wires, both mechanically and electrically.

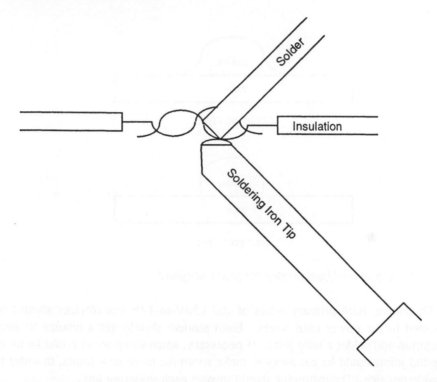

Fig. 7.1 Applying solder to two wires.

It is not really necessary to pre-wet the soldering iron with a small, flat drop of melted solder, but this does help prevent poor joints, because liquid solder transfers the heat from the iron to the wires better than dry iron to dry copper contact would. Similarly, it is not necessary to do all of the above operations in exactly the sequence described, but it is still a good idea.

If there is (1.) *not* enough flux or (2.) *not* enough time for the flux to work, or (3.) *not* enough heat to get the wires up to the same temperature as the melted solder has reached, then any of those three things can prevent reliable joints from being made. The top of Fig. 7.2 shows how a good solder-to-copper boundary should look, with a "low contact angle." An unreliable boundary tends to look like the bottom diagram, with either a 90 degree angle on the right, or a worse one on the left of the diagram.

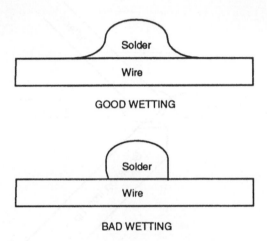

Fig. 7.2 Good and bad solder "contact angles."

Of course, *both* primary wires of the 120V-to-12V transformer should be soldered to the power cord wires. Each student should get a chance to strip insulation and solder a wire joint. If necessary, some scrap wires could be used, or good joints could be cut away to make room for more new joints, in order to provide practice. The instructor should inspect each soldering job.

After the soldering iron has been disconnected and is cooling, each soldered joint should be covered with a spiral of overlapping "electrician's tape," to insulate it for general use, since the transformer will be needed for future experiments in this course. Some of the tape should overlap the wire's own insulation, in addition to overlapping itself on each turn. It is generally a good idea to make two layers of tape, in case it slips and would otherwise expose some bare wire. The taping should also be inspected by the instructor.

Electronic components that might be damaged by heat must be protected during soldering by attaching a temporary "heat sink" such as pair of long-nose pliers. This will be explained on page 160.

EQUIPMENT NOTES

Components For This Chapter	Radio Shack Catalog Number
AC Power Cord, 6 ft.	278-1255 or 278-1253
Wire Stripper	64-2129A
Transformer, 120V to 12V, 450 ma	273-1365A
Soldering Gun, 30 watt	64-2066A
Solder, Flux Core	64-018
Tape, Electrical	64-2348 or 64-2349
Enclosure box	270-1809

Although it is not necessary, it would be a good idea for the instructor of this course to purchase a "ground fault interrupter" socket (see page 57) in a hardware store, attach a power cord (similar to the first item above), and place the socket in a plastic "enclosure" box such as Radio Shack Catalog Number 270-1809. Then all 120 volt ac power, even for the soldering iron, can be obtained via this safety device.

CHAPTER 8

The Oscilloscope

WHAT IT IS

The most important part of an oscilloscope is a large, cone-shaped glass tube with a vacuum inside. It operates the way the old "vacuum tube" radios worked, before the invention of transistors. That is, there is a tungsten wire "filament" at the small end, which is heated by a low voltage but high current source, which could be a battery. Electrons tend to fly off this hot wire, almost like vapor from a boiling liquid. The battery and filament assembly is the left-hand part of Fig. 8.1, on the next page.

Since the electrons are negatively charged, they get strongly attracted to a wire hoop which has a high positive charge on it, from another power source, shown here as thousand-volt battery. (This would work, but in a regular oscilloscope, 120 volt ac from a wall socket is changed into thousand-volt dc by methods that will be explained later in the course.) Some of the electrons get collected by the "anode" hoop and returned to the battery, but many are going so fast that their momentum tends to keep them going to the right, rather than changing direction quickly enough to get caught by the hoop, so they fly through the large hole in the middle. Since the beam of electrons is coming from the *negatively* charged filament (the "cathode" of the 1,000V circuit), this beam is called the "cathode ray." The whole tube is sometimes called a "cathode ray tube," or "CRT." It can be an important part of a TV receiver, in which case it is called the "picture tube."

The electrons continue until they eventually hit the "screen," which has a thin coating of "phosphor" material, such as zinc silicate. This contains impurities that convert the kinetic energy of the electrons into a visible spot of light, and that can be seen outside, through the glass tube. The inside of the tube also has a thin conductive coating on it, to carry the electrons back to the power supply.

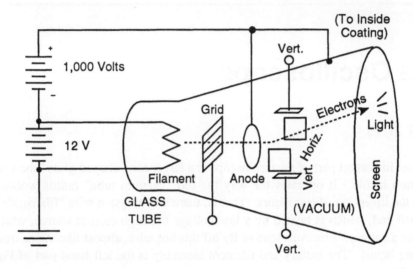

Fig. 8.1 An oscilloscope.

A grid of thin wires is attached to a "terminal" outside the tube (the small white circle), and if a strong negative voltage is applied to the grid, it can stop or at least decrease the intensity of the visible light spot. This is not always used in an oscilloscope, but it is an important part of a TV picture tube.

Four metal plates are on the top, bottom, and sides of the tube, and they are all connected to outside terminals (only two of which are shown in this diagram, for simplicity). If the top one is slightly charged + and the bottom one −, via the "vertical" terminals, the electron beam (cathode ray) is attracted upward, as shown in the diagram, and the spot of light thus appears up high on the screen. Similarly, if the "horizontal" plates are charged by some outside voltage, the spot will move to the side (not shown here).

WHAT IT DOES

Internally Timed Horizontal

If a slowly increasing voltage is put on the horizontal terminals and plates, the visible spot will move slowly across the screen. When the spot gets to the right-hand side, the voltage is arranged to then quickly decrease to zero, so the spot "flies back" to the left-hand starting position. This voltage versus time "signal" that goes to the **horizontal** plates comes from an **internal** "signal generator" which is part of the whole oscilloscope system, and its "waveform" is shown in the diagram of Fig. 8.2.

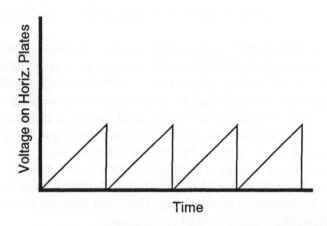

Fig. 8.2 The sawtooth signal that goes to the horizontal plates.

The person using an oscilloscope can control the speed with which the "horiz." voltage builds up, and if it is very slow, the spot of light will only creep across the screen toward the right, then flying back to the left and then creeping across again, over and over. If it builds up and then repeats very fast, a bright line will appear on the screen, since the human eye will merge any motions into a single image, if they repeat faster than about 50 times per second.

Now suppose that another, completely external sawtooth signal generator is being tested, possibly a new one that was built by the experimenter. (In fact, this will be done in a later chapter. Devices that generate unusual waveforms are sometimes called "signal generators" or "function generators.") If that **external** sawtooth signal is fed into the **vertical** terminals, with four times the build-up speed of the **internal** sawtooth signal going to the **horizontal** plates, then the visible spot will trace a path identical to what is shown in Fig. 8.2. As the bright spot moves to the right, it goes up and then suddenly down again, four times before reaching the fly-back spot at the right side.

On the other hand, if a "sine wave" from a 120 volt ac wall socket is fed into the *vertical* terminals, that sine wave shape will appear on the oscilloscope screen. The *horizontal* will still be getting a sawtooth signal from the internal generator, and the *speed* of its voltage build-up (compared to the 60 cycles per second of the standard sine wave) will determine how *wide* the sine wave looks. In other words, if the horizontal sawtooth takes only 1/60 second to build up, before it flies back to zero voltage, then only one complete sine wave will appear, filling the whole screen and looking very large. It will be repeated every 60th of a second, and the human eye will merge these repetitions into a single stationary image.

By comparison, if the internal sawtooth takes a whole second to slowly build up to its maximum value, then 60 little sine waves will appear during that relatively long time, each looking quite narrow. (If the sawtooth speed is not some exact multiple of 1/60 sec., then the image will appear to progressively move, because it is re-appearing at a different place on the screen each time.)

If the *horizontal* sawtooth is *slow*, but a switch is used to *quickly* turn a dc battery voltage on and off at the *vertical* terminals, a "square wave" will appear, which is simply a series of squares or rectangles. Other waveforms can also be made visible with a scope. However, their frequencies (repetition rates) should be exact multiples of the horizontal sawtooth "rep. rate," and therefore the latter has to be precisely adjustable, which it is on all oscilloscopes.

Actually, the terminals from the outside do not go directly to the oscilloscope ("scope") tube. Instead, they go to a variety of adjustable electronic circuits that can make the signals stronger or weaker. Therefore the pattern that can be made visible will fill the whole screen, and it will not be either a tiny thing in a corner that is hard to see, or be too big for the scope screen and partly lost.

The internal circuit that makes a signal stronger (raises the "amplitude" of the signal) is called an "amplifier." It contains at least one adjustable pot, like the one in Fig. 4.3 on page 44, and it also makes use of transistors.

There are *many* adjustments available, and this can be confusing to a new user. Also, the internal circuits ("ckts.") that can make *signals** stronger have a tendency to make "background *noise** " strong also, including radio and TV signals that cause RFI, and also slow-wave EMI (see index) from overhead

---

* The "signal to noise ratio" is often important in electronics.

fluorescent lights. This can make it difficult for a new user to tell which is the real signal being studied, and which is the EMI. The experiments in this chapter will be an attempt to clarify the use of a typical scope, with these problems in mind.

External Signals to the Horizontal and Vertical

Instead of an internal sawtooth signal, the horiz. amplifier can be driven by a signal coming from outside the scope. For example, looking back at Fig. 2.3 on page 18, suppose the *voltage* on a tungsten light bulb could be fed into the *vert.* and the *current* fed into the *horiz.* terminals of a scope. The curve shown in that figure would appear on the scope screen. In fact, we will do this in the experimental section of this chapter. The scope inputs are sensitive to *voltage*, but getting the horiz. to respond to *current* will take some cleverness. (The whole field of electronics is one in which there is much room for creativity, which is probably one reason why the electronics business is growing so fast in the USA, a very receptive country for new ideas.)

EXPERIMENTS

The Basic Oscilloscope

The particular model of oscilloscope being referred to here is Model OS-5020, made by the LG Precision Co., Ltd., of Seoul, Korea, with offices also in Cerritos, CA, USA. Other scopes have similar controls, but some details of the following instructions might differ slightly with various models.

Usually a separable power cord must be plugged into the back of the scope, and then the other end goes into a 120V ac wall socket. A switch marked "Power" is turned on. (On some new equipment, the switch is marked with a circle having a vertical line inside it, which signifies "one or zero," in other words, "on or off.") It takes a few minutes for the CRT to warm up and operate properly.

A long, narrow cable, usually black, has a chrome plated connector on one end and an alligator clip at the other. The chrome connector is called a "BNC" type, but those letters do not stand for anything (the origin is obscure). On the front of the scope are several BNC corresponding connectors, and one is marked "CH2, Y." It has two very small metal pins sticking up and down from its outer cylinder. The BNC on the black cable must be oriented so that *two grooves* in its outer cylinder line up with those *two pins* (usually vertically), and then the cable BNC is pushed firmly onto the BNC on the scope. While still pushing hard, the cable's BNC connector is then rotated about 45 degrees clockwise ("to the right,") in order to lock the two together.

The cable is "coaxial," meaning that it has a central wire inside, and a cylinder of plastic insulation over that, then a wire mesh cylinder over that, and then outer insulation over that, all having the same "axis," or center line. The wire mesh provides three things: (1) "return" conductivity like the neutral wire in Chapter 6, (2) grounding, and (3) shielding against EMI. At the other end of the cable, the wire mesh is connected to an alligator clip, which usually should be attached to the ground of the circuit being tested, since it is permanently grounded inside the scope.

The central wire of the cable is connected to a "test clip," which is opened by sliding a plastic flange (a black disk that sticks out a bit) against a spring, away from the end. This exposes a metal hook, which can be looped around a wire or terminal that is being tested, and of course the flange is then released in order to lock the connection securely. (In some suppliers' catalogs, this type of clip is called a "hook," or "grabber," or "plunger" clip instead of a "test" clip.)

> **Note:**   Be sure that the red button sticking out of the long black plastic handle of the test clip is in the "X1" position (rather than the "X10" position which would divide any input voltages by ten). Similarly, be sure that all buttons on the front panel of the scope itself are on "X1" positions (rather than the "X10" positions which would divide input voltages by ten).

The following are good settings for starting up the scope.

At the end of the cable, attach the grounding alligator clip to the test clip's metal hook, thus short circuiting that cable. The other end of the cable is connected securely to the BNC connector for Channel 2 , Y.

INTENSITY (near the screen):   Middle-range (or lower, counterclockwise).
FOCUS: Middle-range.
    Note: Some buttons and knobs make no difference here, like CH1.
MODE: CH2 (meaning input Channel 2).
POSITION, Y: Middle.
VOLTS / DIV: 5, or a high value (5 volts per large square).
    Note: On the middle of this knob, there is a smaller knob. Turn it all the way
    to the right, until it clicks, indicating its maximum position.

(Table continued on next page.)

(Table continued from previous page.)

AC / GND / DC	DC.
TRIGGER LEVEL:	Middle (up).
TRIGGER MODE:	AUTO.
TRIGGER SOURCE:	LINE.
POSITION, X:	Middle.
VARIABLE:	All the way clockwise, until it clicks.
CAL / VAR:	In (which is *variable,* not calibrating).
TIME / DIV:	1 ms (one millisecond per large division).

A bright line should be visible on the screen. The POSITION, X and POSITION, Y knobs can be rotated slightly, to make the line appear in the middle of the screen. The INTENSITY should be rotated until the brightness of the line is comfortably visible but not extra strong. This slows the aging of the screen, making the scope last longer.

For a short time, the TIME/DIV adjustment can be switched to X-Y, and this will change the line to a spot. That can then be centered, and the TIME/DIV should then be switched back to 1 ms, in order to prevent burning out the screen in the middle.

The Sine Wave

Using a 120V-to-12V transformer with a power cord soldered to its primary coil, make sure that the bare metal tips of the secondary wires are not touching each other, and then hook the two yellow wires to the BNC cable's test clip (hook) and alligator clip (ground). Since this will be an ac experiment, and since the transformer's secondary is not grounded, either yellow wire can be attached to the scope's ground. Plug the transformer's power cord into a wall socket (preferably via a ground fault interrupter).

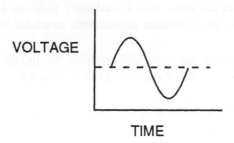

VOLTAGE

TIME

Fig. 8.3 A sine wave, as seen on the oscilloscope screen.

Part of a sine wave will appear on the screen. The TIME/DIV adjustment has a fine tuning knob labeled VARIABLE, and this can be rotated to the left until the sine wave is complete, as in Fig. 8.3 on the previous page. The small VARIABLE knob mounted in the middle of the VOLTS/DIV larger knob can also be rotated, until the height of the wave decreases to a level of better visibility.

The full "cycle," as shown in Fig. 8.3, takes 17 milliseconds ("ms"), so there are 60 cycles per second ("cps"). Instead of using the old term, cps, modern publications use the term "Hertz," abbreviated as "Hz," which means the "frequency" in cycles per second. (In many European countries, the standard ac voltage is 240V, with a frequency of 50 Hz.)

If there is a TRIGGER SLOPE button, change its position. The sine wave will be seen to flip upside down. The "trigger" is a feature which starts the bright spot moving at a certain time, in this case whenever the power line voltage starts to move from zero. If the button was set previously to +, then the spot started to move when a positive voltage was sensed. Changing the button position causes the spot to start moving at the opposite input voltage. Now rotate the TRIGGER LEVEL knob to the right, and then to the left. The sine wave moves horizontally, because the beginning of the spot movement (that is, the triggering) has been set at a higher or lower voltage.

Now change the TRIGGER MODE from AUTO (automatic) to NORM (normal), and change the TRIGGER SOURCE from LINE (the 120 volt ac power line) to VERT (the vertical input, which in this case is CH2/Y). This frees the "trigger" for optional adjustments, instead of being automatically tied to the timing of the ac power line. Rotate the TRIGGER LEVEL knob a very small amount, until the sine wave appears. Then rotate the VARIABLE knob (the fine tuning of the TIME/DIV setting) a small amount, until the sine wave moves horizontally. You might be able to fiddle with the TRIGGER LEVEL and VARIABLE knobs to get the wave into a stationary position again, but it is difficult to do this without the automatic adjustments available in the "AUTO" mode.

The waveform displayed on the scope will probably be slightly distorted from the perfect shape shown in Fig. 8.3. Quite often the ac power line delivers badly shaped waves, and the small transformer might also cause further distortion due to nonlinearities in the characteristics of its iron core, including "saturation," which will be discussed later in the section on magnetic amplifiers.

Disconnect the BNC coaxial cable's hook from the transformer, and then pull out the power line plug of the transformer. Note that the cable is disconnected first, so that an inductive kick (Chapter 1) will not damage the oscilloscope.

(In most cases, there probably would not be any damage, but that depends on how fast the plug is pulled.)

The Square Wave

Keeping the other scope settings intact, change the TIME/DIV to 10 milliseconds. Slightly rotate the TRIGGER LEVEL knob until the bright spot moves repeatedly across the screen (probably at its "zero" setting, straight upwards). Attach the black grounded alligator clip of the BNC coax cable to the negative terminal of a 9 volt battery. Repeatedly touch the hook of the cable to the positive terminal, for a very short time and then take it away. "Square" (really rectangular) waves will appear on the screen. The TIME/DIV knob might have to be adjusted to 20 ms, or else the VARIABLE knob's setting might have to be changed, or other settings such as the X or Y positions could possibly be optimized, in order to provide better looking waves.

High Input Resistance and EMI

Keeping the same settings, have one end of a clip lead attached to the hook, but the other end not attached to anything. Have the black grounded BNC's alligator clip attached to the battery or else to another clip lead, going nowhere. Turn the VOLTS/DIV to the maximum sensitivity setting, 5 millivolts (mV). Badly distorted sine waves will march across the screen. If the TRIGGER MODE is set to Auto again, and the TRIGGER SOURCE to Line, the TIME/DIV can probably be changed slightly in order to make the waves stand still (really repeat at the same place).

The sine wave "signal" being picked up by the two wires comes from electromagnetic waves generated by wires in the walls, from the "ballast" inductors of fluorescent lights, and even from the power cord of the scope itself. This is typical EMI, and it can play havoc with weak signals that are expected to be useful, such as radio waves from distant sources. As was mentioned in an earlier chapter, the two clip leads act as a radio antenna (aerial), picking up a voltage from the electromagnetic waves in the air. If the small transformer is plugged in again, with its yellow wires loose and near the hook, the waves will get *much* bigger, and the VOLTS/DIV can be de-sensitized to about 50 mv. The same thing happens if the soldering iron is plugged in and placed nearby, but the effect is not as large, because the iron only has a small coil inside.

Now "short circuit" the hook by connecting it directly to the alligator clip of the BNC coax cable. Naturally, the signal disappears. Instead of nearly zero ohms of a "short," attach the hook and the alligator to the ends of a resistor. A 5K pot will cut the signal down somewhat, and 100 ohms (or a 5K turned down most of the way) will further reduce the EMI. Although a general rule on page 51 called for "high input resistance" in order to prevent dragging down a driving voltage, it has the disadvantage of tending to pick up EMI. Therefore, the optimum input resistance depends on the situation, but some kind of compromise might be necessary to "get between the horns of the dilemma," and avoid both problems. (With the "twisted pair," and "balanced lines," and/or "shielding," as described in the next sections, high input resistance *can* ordinarily be used.)

The Twisted Pair

Twist two clip leads around each other, and use them as the antenna, attached to the BNC coax hook and the grounded alligator, as above (not "shorted" to each other at the end). This "twisted pair" does not pick up as much EMI as two straight wires do. For one thing, the clip lead that goes to the grounded alligator is acting as "shielding," as previously discussed at the bottom of page 57. In addition, even if there were no ground connection to that wire, any electromagnetic wave that hits one wire also hits the other, and the two voltages that are thus generated tend to cancel each other. (Remember that the scope only detects a *difference* in voltages between the hook and the alligator.) Therefore, twisted pairs are used widely in electronics to prevent strong EMI. In electronics diagrams, instead of trying to draw wires twisted together, the twisted pair is symbolized as in Fig. 8.4, where the parallel wires are not drawn as being twisted, but instead a figure-eight is drawn over them, indicating to the reader that the wires themselves are actually twisted, not really parallel.

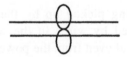

Fig. 8.4 Electronics symbol for a twisted pair.

Balanced Lines

Under some circumstances it is better *not* to ground *either* current-carrying wire, but make a twisted pair out of them instead. Telephone circuits are often wired this way. An electromagnetic wave (from a power cord, from radio, or from any other source of EMI) will hit both wires at close to the same time. Whatever device is attached to the ends of the two-wire cable will probably respond to a *difference* in voltage between the two wires, and in fact an earphone does exactly that. (If one wire was grounded, then the earphone would respond to the voltage compared to ground, but if neither is grounded, then any kind of voltage difference will do.) This is called a "balanced line," because EMI voltage tends to appear on both wires equally, in a balanced fashion. However, if the wires are not twisted around each other, they are less likely to get equal EMI and balance out.

There is another reason for using the balanced line type of wiring: the type of line having one of the conductors be a "common ground" can cause the "ground loops" that were discussed on page 61, unless all the parts of the system are close together and very well grounded. The "close together" situation is not possible with telephones, so ground connections tend to be avoided in that application.

Shielding

The hook wire of the twisted pair experiment on the previous page would be better protected against EMI if the "shield" clip lead covered its wire completely, instead of being on just half or even less of its area. The BNC's coax cable does this very effectively, since its outer conductor (not visible, but attached to the alligator clip and the ground) is a woven mesh of copper filaments, going all around the inside wire. Disconnecting the clip leads and taking the BNC cable's own unshielded black alligator clip *far away* from the little 12 volt transformer just about kills the EMI completely, because the magnetic waves from the transformer mostly hit the well-shielded middle of the cable.

Twisted pair balanced lines work better when there is a *third* conductor around them, a separate shield cylinder of wire mesh, which is thoroughly grounded. To save money, telephone wires are usually just twisted but not shielded, although newer ones designed for high performance internet use, etc., are often being shielded by a third conductor, which is a grounded mesh.

In stereo/audio equipment designed for "professional" use (in recording studios, as compared to homes), "balanced lines" (pairs of wires) are ordinarily used instead of the "unbalanced" patch cables supplied with home amplifiers, etc. The balanced line is usually twisted and then covered with a third copper mesh shield that is grounded. At both ends of the cable, there is a fairly large "XLR" type of connector, with three pins or sockets, instead of the home-use "RCA" connector with only two (just inner and outer, and no twisted pair).

The symbol for shielding is shown in Fig. 8.5 (next page). Although it is actually a very long cylinder of wire mesh, it is drawn as a small loop, for simplicity. Sometimes the loop is drawn as a solid line, and sometimes as a complete cylinder around the pair of wires. It can also be drawn around a single wire. (The figure-eight in the twisted pair illustration of Fig. 8.4 can be combined with the loop of the grounded shield in Fig. 8.5, all on the same drawing.)

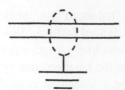

Fig. 8.5 Electronics symbol for a grounded shield.

The bottom wire of the pair might or might not be grounded (not grounded if a balanced line). The long shield might or might not be connected to each chassis at the ends of the wires, although it usually is connected to both. The pair of wires might or might not be twisted around each other.

In many electronic systems such as stereo/audio systems for home use, there is often just one wire inside of a common-ground shield. It is difficult to avoid ground loops with such wiring. These situations can result in destructive "feedback" oscillations (audible howling or repeated loud thumps). One way to stop this if it begins, is to disconnect the grounds of one component, such as a subwoofer, but keep the ground to the other component such as an amplifier. The common-ground pair from the amplifier goes to the primary of a 1:1 "isolation transformer" (Radio Shack catalog number 270-054), which does not change the voltage. The secondary of the transformer can be either grounded or ungrounded (both can be tried). Isolation transformers are used in other applications also, where grounds must be avoided in the secondary circuit. (See also page 61.)

X Versus Y Inputs
In the next experiment, the TIME/DIVN knob is rotated all the way to the right, and that selects the X versus Y type of input. There are no longer any sawtooth waves going to the horizontal plates, but instead, the horiz. input is another BNC connector, the one labeled "CH1, X." The experimenter should now plug in a second coax cable into *that* BNC connector on the scope, keeping the first coax in CH2, Y. (The MODE is still on CH2, meaning that the Y input is still going into that channel.) The two alligator clips on the BNC cables are then attached to each other, although they are both grounded and already connected internally. This just makes a better ground.

A single bright spot appears on the screen, and it can be centered with the POSITION X and Y knobs. The INTENSITY should be turned down if the spot appears extremely bright.

The two VOLTS/DIV knobs are both turned all the way clockwise, to 5 mV, for maximum sensitivity. A diagonal line appears on the screen. (If it is very small, plug in the 120V-to-12V transformer and place that nearby, with nothing attached to its secondary, and none of those "secy." wires touching each other.)

If a separate clip lead is now attached to each hook, the line gets bigger. (Possibly it will be somewhat oval-shaped, which would be due to "phase differences," to be discussed in a later chapter on capacitors.) If the diagonal line is too tall for the screen size, decrease the sensitivity of the VOLTS/DIV for the CH2, Y cable to about 20 mV, or whatever makes the line go to about a 45 degree angle, all within the screen. If it is too wide, decrease the CH1, X sensitivity.

Now place the clip leads close to each other, in parallel positions, but take your hands away from them. The line or oval should get smaller, because there is more of a balanced line, where any magnetic wave will hit both wires equally. Twist the clip leads together (without the metal alligators touching), and the line gets still smaller, because of the twisted pair effect.

Attach the two hooks to a 5K pot, arranged as a rheostat (page 46), and notice that the line or oval gets smaller as the resistance decreases, confirming the effect of a low input resistance, as discussed on page 79. Attach the two hooks together, and note that the line gets even smaller. However, it probably will not collapse to a tiny spot, because of slight leakage of magnetic waves into the hooks and alligator clips — their resistances are not quite zero.

The Curve Tracer

The voltage versus current "characteristic curve" of the tungsten light bulb (mentioned on page 75) can be directly obtained with the X versus Y inputs of a scope. The experimenter should construct the circuit of Fig. 8.6. The horizontal

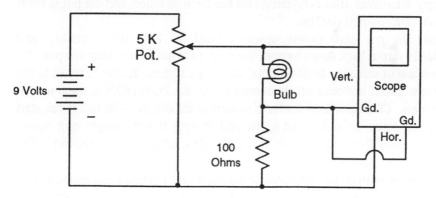

Fig. 8.6 A curve tracer, measuring V versus I for a tungsten bulb.

BNC is CH1,X and the vertical is CH2,Y. This time, the negative terminal of the battery is not connected to the grounds. Instead, the negative wire goes to the horizontal, and both grounds (alligator clips) are connected to the bulb.

The *voltage* across the bulb at any given time is shown on the scope screen as the vertical displacement from zero. But how do we measure the *current*, since inside the scope tube, the horiz. plates will only respond to voltage? The answer is to use good old Ohm's law, in a clever way, where the current through a 100 ohm resistor always must be associated with a voltage across it. The characteristic curve of the 100 ohm ordinary resistor is quite linear, as on the left side of Fig. 2.3 (page 18), so we can use that resistor as a "sensor," translating current into voltage.

There is a good reason why the negative wire from the battery is not connected to "ground" this time. The outer conductors of both BNC connectors are permanently grounded, for safety reasons, to both the neutral and the green wires of the 120 volt ac wall socket. Therefore, the alligator clips of both BNC cables have to be used for something that can be "common" to both the *voltage* sensor (the vertical input pair of wires) and the *current* sensor (the horiz. pair of wires.) Since the negative wire from the battery can't go to that common point, it can't be grounded — fortunately, there is no reason why it has to be grounded.

Do not connect the battery at first. Set the VOLTS/DIV of the horizontal (X) to minimum sensitivity, 5 volts/divn., and the vertical (Y) to 1 volt/divn, with both VARIABLE knobs fully clockwise. The TIME/DIV is set on "X - Y." The 100 ohm resistor should be a 1 watt or higher type, although 1/2 watt will do if the times of usage are kept short, but using a 1/4 watt is not a good idea. Connect the battery + terminal after everything else has been attached, and the pot is set at its bottom (zero output) position.

Rotate the pot shaft slowly upwards until the bulb lights visibly, and immediately turn it back down most of the way, to avoid overheating the pot. Do this again and observe the bright spot on the scope screen. It should move to the left or right, and if it moves off the screen, re-set the POSITION X knob until it stays in view. (The spot does not have to start at the center — in fact it can start at a corner.) Raise the pot output again, and observe that the bright spot moves horizontally and then starts to go vertically also, after about a second. The vertical motion corresponds to the bulb filament heating up and the resistance rising. (One or both of the VARIABLE knobs might have to be changed to less sensitive settings, in order to make these changes in position easily visible.) To make the characteristic curve be flipped so it is oriented like on page 18, disconnect the battery and reverse its + and − connections, and also use the button marked "NORM versus INV" to invert the scope display.

OPTIONAL

If the 12V ac secondary of the transformer is substituted instead of the 9V dc battery-and-pot as a power source, then the bright spot becomes a diagonal, continuous line, because the increase and decrease of voltage becomes too fast (60 times per second) for the eye to distinguish as separate images. Also, it goes both up and down from zero, since the ac goes to both + and − from zero volts. However, the visible image is now a *straight line,* not the curve of Fig. 2.3, because each "cycle" of current goes too fast for the tungsten filament to cool off when zero current is reached, so the lamp stays hot, and the resistance does not decrease enough to be seen on the scope display. (Even with dc, a residual heating effect is observable.)

To remove the BNC cables, push the BNC in against its spring, then while holding it in, turn it 45 degrees to the left, and pull the connector out. The VOLTS/DIV knobs should be turned to the left before turning off the scope.

OPTIONAL

The reader might be interested to know that the incandescent electric light, although generally credited to Thomas Alva Edison because of his U.S. Patent of 1879, had previously been patented by J. W. Starr in 1845 * and again by M. G. Farmer in 1858, all using carbon filaments in vacuum tubes. However, efficient vacuum pumps and electric generators had not yet been developed by other inventors until Edison's time. He was "at the right place at the right time," but also multitalented and extremely energetic (Edison only slept about 4 hours per night). Besides being a technical genius, Edison had extraordinary abilities in getting venture capital from Wall Street, and in organizing the companies that later became General Electric and Consolidated Edison, to make and distribute the necessary electric power for the widespread use of electric light bulbs. ("Successful engineering is not just engineering.")

TELEVISION

On page 73, the vertical input had a fast sawtooth signal, and the horizontal had a slow one, so the repeating-triangles pattern would have appeared on the screen display. Now the reader is asked to imagine a horizontal

---

* Popular Electronics, March 1998, page 45.

sawtooth that is moving the bright spot across the screen so fast that it goes across 483 times in only 1/30 of a second (or 483 X 30 = 14,490 "scans" per second). That would look like a bright *horizontal* line. (The fly-back is *much* faster — too fast to be visible.)

However, imagine a slower *vertical* sawtooth scan rate of 1/30 second, starting whenever the first horizontal scan starts. That would make each successive horiz. scan line appear a little bit higher on the screen, and by the time the 483rd line was made, the whole screen would have been filled in with white, or whatever color the spot is — yellowish green for most scopes. (Then the "scanning" would start all over again, at the lower left.) This filled-in screen full of lines (really a single moving dot) is called a "raster." The human eye has "persistence of vision," which is a kind of slowness of response that makes the whole screen look like a continuously bright square.

That is what occurs in the picture tube of a TV receiver. However, a negative voltage on the "grid" electrode shown on page 72 makes the spot less bright in certain places, creating a picture just like a photograph. The "resolution" is such that a single line consists of 330 dots, in a horizontal array.

Like most things in modern electronics, it is slightly more complicated in reality. For one thing, there are actually 525 horiz. scans in each 1/30 second, but 42 of those scans occur with a strong negative grid voltage, so they are not visible and are used for other purposes. Another complication is that the next time that the 483 are scanned, the whole raster is a tiny bit higher, so those lines fall between the previous lines, thus "interlacing" the two rasters. The time after that, the lines go back to their original positions. This helps fill in the dark spaces between lines, which might otherwise be visible as black lines. But none of that alters the basic idea of drawing a picture by making closely spaced parallel lines (once again, really just fast-moving bright dots).

The signals from the TV station to the receiver are sent via *very high frequency* ("VHF") radio waves. These can have frequencies anywhere from 54 to 216 *megahertz* ("MHz"), or million cycles per second, depending on which "channel" is being transmitted. Even higher frequencies are used for cable and satellite TV, and common "downlink" frequencies for the latter are around 4 *gigahertz* ("GHz"), or 4 billion cycles per second (see page 212).

At the TV studio and transmitter (or in a hand-held "camcorder"), there is "vidicon tube," somewhat similar to the scope tube on page 72. In this case, however, light comes from an object in the studio and goes through a lens that focuses it onto the face of the tube. Instead of the phosphor on page 72, the inside of the screen is covered with a "photoconductive" material, such as cadmium selenide, spread into a square array of closely spaced dots. When light hits this material, it conducts electricity, but not if there is no light shining on it. There is also an extremely thin layer of electrically conductive material on the inside of the tube face. (This can be the transparent semiconductor, indium oxide doped with tin oxide, "ITO.")

The sawtooth signals on the horizontal and vertical plates are timed to exactly coincide ("synchronize") with those in the TV receiver. As the beam of electrons inside the tube scans across the screen, if light is hitting that particular place, the photoconductor allows the electricity to go through it and put a charge on the thin conductive layer, which is connected to the TV transmitter, sending a signal via VHF radio waves.

At that very moment, the receiver's electron beam is hitting a corresponding place on the screen inside the picture tube. The incoming TV signal tells the receiver to put a slight positive charge on the grid and thus let the beam go through and show a bright spot.

However, an instant later, if the vidicon tube's electron beam scans to a spot where there is *no* light from the camera lens, then the TV signal tells the receiver to put a strong *negative* charge on the grid, *cutting off* the beam that would have made a bright spot, and therefore that part of the picture tube looks *dark*. Thus the light spots at the TV studio cause light spots in the TV receiver, and dark spots lead to dark spots.

EQUIPMENT NOTES

Components For This Chapter	Radio Shack Catalog Number
Battery, nine volts	23-653
Clip leads, 14 inch, one set.	278-1156C
Lamp bulb, tungsten, 12 volt dc, green or blue	272-337A
Transformer, 120V to 12V, 450 ma, center tapped	273-1365A
Resistor, 100 ohm, 10 or 1 watt	271-135 or 271-152

Oscilloscope, Model OS-5020, made by LG Precision Co., Ltd., of Seoul, Korea, with offices also in Cerritos, CA, USA (or similar inexpensive scope).

CHAPTER 9

Capacitors

WATER ANALOG

Imagine a system of water pipes as shown in Fig. 9.1, with a wide cylinder, and a movable piston inside that. A spring tends to keep the piston near the middle. A rubber O-ring around it (shown in cross section as two black dots) makes a fairly good seal, but the piston can still slide to the right or left if water pushes it.

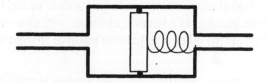

Fig. 9.1   A water analog of a capacitor.

Water coming into the cylinder will push the piston until the restoring force of the spring becomes equal to the force from the water, and then all motion will stop. Therefore water that moves toward the right, as shown in the diagram, will only be able to move for a short time. The larger the *size* (the volume) of the cylinder, the more *mass* of water will be stored in it before it stops, with constant pressure driving the system. Also, if the pressure is changed, but the size is held constant, then the more water *pressure*, the more *mass* of water will be stored in the piston/spring system before it stops "charging up." (This would be similar to pumping air into a tire — the more volume and/or the more pressure, the more total mass of air will be put in, until the inside pressure becomes equal to the pumping pressure, and then it will stop.) Therefore, we could write an equation where

$$\text{(mass moved in)} = \text{(size) (pressure)} \qquad (9.1)$$

It is important to realize that water being pushed *in just one direction* (to the right, for example) will *eventually* be stopped, whenever the cylinder pressure becomes equal and opposite to the driving pressure (analogous to *dc electricity*). If the cylinder is small, this will happen pretty fast. However, water that is only moving *short* distances back and forth will be able to continue its "alternating" flow (analogous to *ac electricity*), whether the piston is in the pipe or not, provided those motions are indeed very short. Longer motions could go back and forth, if the cylinder is *larger*, although each motion will probably take a *longer time* before it reaches its end point (analogous to ac electricity of *low frequency*).

These relationships are all close analogs to the behavior of electricity going into a capacitor, where mass of water is like coulombs of electric charge, size of the cylinder is like "capacitance," and pressure is like voltage. A bigger capacitor will allow longer-time back and forth alternations of electricity.*

WHAT A CAPACITOR IS

A capacitor (called a "condenser" in older literature) consists of *metal plates*, separated by a thin layer of *insulation*, which could be vacuum or air, but usually is plastic or ceramic. When attached to a battery, electrons will flow onto one metal plate, and they are pulled off the other one, leaving positive charges there.

COMMENT

As the reader possibly knows, "electricity" in metal wires is entirely the flow of electrons, but protons or nuclei do not move very far through metal. However, it is convenient for us to *imagine* that *either* electrons are going in one direction *or* positive charges of some mysterious kind are going in the opposite direction, although not both at the same time in the same place (that would double the current). This is what the early scientific researchers believed, before electrons were discovered, and it is still a useful idea, even though only the first half of it is true. Sometimes the mathematical relationships in electronics, left over from the old days, are still written in terms of positive charges moving. This is especially found in older books about battery science, electroplating, and electric motors.

---

* A small capacitor allows a small ac current. Thus the human body conduction on page 6.

When the charge accumulates on the metal plates of a capacitor (see top of Fig. 9.2), a *voltage* appears, and this is equal to the total charge divided by the distance between the plates, as previously described in the dimensions at the bottom of page 15. If there is just vacuum between the plates, this situation is shown at the top of the figure. The meter is a high quality multimeter, which can measure the capacitance directly. (The meters used in this course do not, unless the instructor is able to obtain a special one for demonstrations.) It determines the

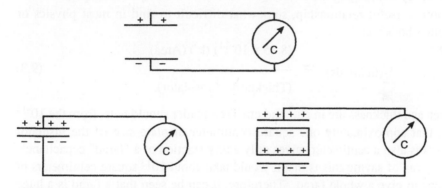

Fig. 9.2 Capacitors, with vacuum or ceramic between the plates.

amount of charge that flows into the capacitor when a certain voltage is applied from within the meter. When the voltage on the capacitor builds up to be equal to the driving voltage coming from the meter, the charging process stops. (Another useful analogy is to consider the capacitor to be like a very small storage battery, which will stop charging when its voltage becomes equal to the source voltage.)

An important observation is that more charge can be put on the capacitor, at a given voltage, if the metal plates are closer together, as at the lower left side of Fig. 9.2. The reason for this (which quite surprisingly is not usually taught in physics or electronics courses) is that repulsion of the accumulated charges is what eventually stops further charges from coming onto the plates, but the electric field (page 15) from the positive plate neutralizes *some* of the field on the negative plate, and vice versa. This allows more charge to come in, before the flow finally stops. The closer the plates are together, the more neutralization occurs, and the more additional charge can come in.

Capacitance, C, is the amount of charge, Q, *per voltage, V,* as in the following equation .

$$C_{(in\ farads)} = \frac{Q_{(in\ coulombs)}}{V_{(in\ volts)}}$$ (9.2)

Therefore, by rearranging the terms, $Q = CV$, similar to eqn. 9.1 on page 90.

Another useful relationship, somewhat difficult to find in most physics or electronics books, is

$$C_{(in\ farads)} = \frac{(8.85 \times 10^{-14})\ (k')\ (Area)}{(Thickness\ of\ insulator)}.$$ (9.3)

The area and thickness are in cm^2 and cm. The reader should note, from the 10^{-14} factor, that approximately one square centimeter of plate size (if the insulator thickness is also a centimeter) gives only a tiny fraction of a "farad" capacitance. Another way of saying this is that it would take about 10^{13} square centimeters of plate size to give a whole farad. Therefore, it can be seen that a farad is a huge quantity of capacitance.

Most practical-sized capacitors have values that are only in the range of microfarads, usually referred to as "mfd." Sometimes the values are in picofarads ("pf," pronounced "puff," and equivalent to 10^{-12} farad). It would be useful if the k' factor could somehow be made to be a large number, to make up for the very small 10^{-14}.

That k' factor is called the "relative dielectric constant," and it is dimensionless. If the insulator (which is referred to as the "dielectric" in physics) is a vacuum, then $k' = 1$. Common solid insulating materials like plastic or mica have k' values ranging from about 2 to 10. Therefore, putting them in between the plates instead of a vacuum raises the capacitance. This is shown at the lower right of Fig. 9.2, where the rectangle of solid dielectric allows more charge to accumulate on the plates, just like putting the plates closer together would have done. It should be noted that k' is the capacitance with a certain dielectric, *divided by* the capacitance with only a vacuum between the plates, and that is the reason for the word "relative."

How does the dielectric material raise the capacitance? (Various other math is usually given in textbooks and courses, but again, the reasons "why" are usually not explained.) All solid materials consist of positively charged atomic nuclei, with electrons going around them in orbits. In the strong electric field caused by the charged metal plates, the electrons of the dielectric tend to "polarize," which means that they are attracted somewhat toward the positively charged plate (toward the top, in the lower-right diagram of Fig. 9.2). The more they can move, the more they polarize, and the higher their k' value is. In some materials they are more free than in others, although they all *stop* after going a few Angstroms of distance, because they are insulators. (If the charges *continued* to move, then these materials would be *conductors,* not insulators.) When the electrons can spend more time than usual up near the positive plate, their positively-charged nuclei (shown at the bottom of the rectangle in the figure) are exposed more to the negatively charged bottom plate. These positive fields tend to neutralize the negative charges on that plate, thus making *less repulsion* of new electrons, which can then move in. The more charges moving in at a certain voltage, the more capacitance, according to equation 9.2. With some modern dielectrics such as barium titanate or lead magnesium niobate ceramics,* the dielectric constant raises the capacitance 1,000 or even 20,000 times greater than with vacuum.

Why not just make the spacing extremely small? Because the electricity can jump spontaneously across very small gaps, making sparks. Thin materials such as mica that can be put in between the plates and prevent sparks are said to have high "dielectric strength," which is not the same as having a high dielectric constant. Instead, it means that a very high voltage can be put across the insulator without causing "dielectric breakdown," in which a spark would "punch through" the material. The goal of modern capacitor design is to make the insulator be as thin as possible with a high dielectric *strength,* and also to have as high a dielectric *constant* as possible. A great deal of progress has been made in this technology during the last few years. (For examples, see pages 266 and 267.)

Some capacitors are made by separating *two* very long strips of aluminum foil by a very *thin* sheet of plastic insulator. This sandwich is then rolled up into a convenient cylindrical shape, having a fairly large area (see eqn. 9.3) and a fairly small thickness.

It should be intuitive that having a large area is equivalent to having many smaller-area capacitors hooked up in parallel, and this is quite true. A practical

---

* Ceramics are made by sintering. More description is in the Glossary, page 274.

way to have many in parallel is to use the *multilayer* configuration of Fig. 9.3. The black lines are metal layers, and insulating layers (not indicated specifically in the diagram) lie between them. The more insulator layers (four in this diagram), the more capacitance. Up to 100 layers are used in modern devices, also combined with ceramic dielectrics having high k' values.

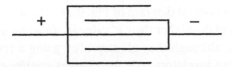

Fig. 9.3 A multilayer device, where the capacitance is multiplied by four.

Just like pumping up a tire with compressed air, a capacitor will charge at the fastest "rate" in the beginning, as shown at the left side of Fig. 9.4. The rate is the slope of the curve at any given time point, and it is proportional to the current. As the *capacitor voltage* gets closer to the *source voltage,* of course the difference gets vanishingly small, and it is that voltage difference that drives the current. Therefore the current versus time curve (not shown) would look like the "discharging" curve at the right — high in the beginning and low later on. Theoretically, the end point would never quite be reached, but actually the current soon gets down into the random "noise" level, where it can no longer be measured and is effectively zero. The mathematical description is an "exponential" equation, similar to compound interest financial equations, or radioactive decay equations. These equations would probably not be useful to readers of this book, so they are only described briefly in the footnote.* (The Greek letter is explained on the next page.)

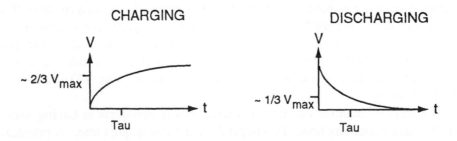

Fig. 9.4 The voltage across a capacitor, versus time.

---

* When charging, $V = V_{max} (1 - e^{-t/\tau})$. When discharging, $V = V_{max} (e^{-t/\tau})$. $I = V / R$.

During discharging at the right side of the figure, the time it would take to get to zero voltage would be difficult to determine. Therefore, a more convenient *discharging time* is defined as the time it takes for the voltage to get to about 1/3 of the original maximum value. The ~ ("tilde") symbol means "approximately," and the exact fraction is 1/2.718, where the 2.718 is e, the base of the natural logarithm. This calculates to be about 37%, which is very roughly 1/3. This standard discharge time is usually called the *time constant,* and its symbol is the Greek letter τ ("tau").

During charging, the time to the full V_{max} is theoretically infinite, as mentioned on the previous page. However, the more convenient *charging time* is the time it takes to get up to $(1 - 0.37)V_{max}$, or about 63% of V_{max} , and we can call that ~2/3 of the V_{max} . Again, this is called the *time constant*, τ.

A useful relationship is that the time constant is given by

$$\tau = RC \tag{9.4}$$

where R is whatever resistance is slowing down the charging or discharging, and τ is in seconds. In Fig. 9.5, it looks like that R is about 2,500 ohms, both for charging and for discharging. If the capacitance happens to be 1,000 mfd, that is the same as 0.001 farad. Then the time constant, τ, for either charging or discharging is $(2,500)(0.001) = 2.5$ seconds.

EXPERIMENTS
Voltage Versus Time
The circuit of Fig. 9.5 should be constructed, where the "SPDT" switch is just a

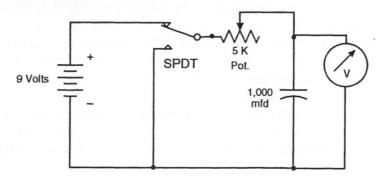

Fig. 9.5 Circuit for charging and discharging a capacitor.

clip lead, with only one end attached to the 5K pot, and other end not being clipped onto anything. That loose end will later be touched to the battery's + terminal, and then later still, to the − terminal or any wire attached to it, which will make it equivalent to a *single pole, double throw* ("SPDT") switch. The term "double throw" means two possible positions, up and down as shown here.

If it was a DPDT type (*double* pole, double throw), then there would be two switches hooked together mechanically, so two movable parts would simultaneously make contact with two upper terminals, or else with two lower terminals. Thus there would be 6 terminals altogether. At any rate, we will use no real switch at all, just a loose clip lead.

Note the symbol for a capacitor, with a curved bottom line. Older literature used two straight lines, such as the one shown in the lower left of Fig. 9.2. However, that is the same as the symbol used by electricians to mean the contacts of a relay or switch, as will be discussed later in Chapter 12. To prevent misunderstandings, electronics engineers have changed their capacitor symbol to the one with a curved bottom. We will use a "polarized electrolytic" type with a capacitance of 0.001 farad, which is unusually large. The white arrow on the case points to the negative terminal, and the capacitor might be damaged if that wire is made positive for a long time. (The insulator inside is a very thin layer of aluminum oxide on a rolled up strip of aluminum foil. This oxide is generated electrolytically, and reversing the voltage could remove it.)

Note also that if "leads" (wires coming out of a device such as a capacitor) have to be bent, they should come out straight for a short distance and then only be bent at a place several millimeters away from the device itself. On the other hand, if they are bent right at the edge of the device, then the insulation around the device is likely to become damaged.

Adjusting the 5K pot (used here as a rheostat) to its highest resistance, and setting the multimeter on the 15 volts dc range, touch (or actually clip) the loose clip lead to the battery. The voltmeter will take a few seconds to reach the final 9 volt reading, following a curve like Fig. 9.4. Then touch the previously loose clip to a negative wire, and observe the discharge behavior. Turn the pot to a lower resistance setting, but **not all the way to zero**. Go through the same procedure and observe the faster time constant.

The reasons for not using the minimum resistance setting are that (1) too much current would flow, possibly damaging the capacitor, and (2) all that current would go through a tiny area of resistor material inside the pot, overheating and probably destroying it. *If* the circuit was using that device as a *true potentiometer*, then during *charging*, the pot should *not* be set at the *maximum* voltage, because too much current would pass through a small area.

Voltage Versus Current

Set up the circuit of Fig. 9.6, which is the same as the one on page 83, except that the capacitor has been substituted instead of the light bulb. The arrow on the outside of the capacitor should point to the ground, since that is negative. The scope settings can be the same as on page 84, but the sensitivities can be changed somewhat if desired, because the voltages and currents do not have to be so high this time. If the battery polarity is reversed (and the NORM / INV button is pressed), then the capacitor leads can also be reversed, although using it the wrong way for a short time will not really hurt it.

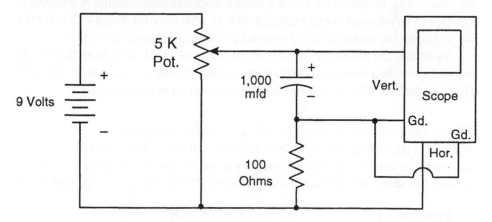

Fig. 9.6 Voltage versus current in a capacitor.

A curve similar to the one for the light bulb is observed, except that when the pot is turned down again, a reverse current flows, while the capacitor ("cap") is discharging, and this did not happen with the bulb. At any rate, the current changes (either increasing or decreasing) *first*, while the cap is charging or discharging, and then the voltage changes a short time *later*. Going up and down quickly with the pot makes a complete *oval* pattern on the scope screen.

SAFETY NOTE: Whenever *anything* is plugged into the 120 V ac wall socket, as it will be in the next experiment, it would be best to have it go through a ground fault interrupter. Even the small 12 V transformer could possibly have an internal short circuit from a primary wire to the metal case, or to one of the secondary wires. Although that seems unlikely, if thousands of people read this book and do the experiments, it *could* happen to one person. If a GFI is not available, then the instructor might use a neon tester, having one hand on one wire and touching the other wire (page 5) to each 12 V transformer's metal case and secondary wire in the class, while the power cords are plugged in. In addition, and more important, **students should observe the one-hand rule** (page 4) when using 120 volts.

In this experiment, plug in the transformer and use its 12 V secondary instead of the battery. With the X versus Y type of scope display, a continuous oval pattern is observed, which is called a "Lissajous figure," named after an early scientist. The vertical VOLTS/DIV knobs, and also their smaller VARIABLE knobs, can be adjusted to other settings such as >5V / div for X and <20 mV / div for Y, in order to make the scope pattern become nearly a perfect circle.

Switching the MODE from CH2 to DUAL and TIME/DIV from X-Y to 20 ms will provide a two-wave display, much like the right-hand side of Fig. 9.7. (Unplug the BNCs *before* the transformer, to avoid an inductive kick.)

Phase *(No experiments, except for the optional one on page 100)*
Why does the voltage build up later than the current? The answer is possibly known to the reader, but a lot of thought should be devoted to this point, since it is useful in many areas of electronics. At the left of Fig. 9.7 is a pendulum, in

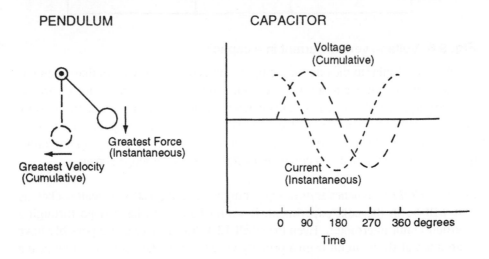

Fig. 9.7   "Out of phase" situations in a pendulum and a capacitor.

which the velocity is *out of phase* with (*does not reach its maximum at the same time* as) the force that causes it. At first sight, one might expect that, at the very instant when the downward force is greatest (at the position farthest to the right), the velocity should be greatest, and when the gravitational force is zero, then the velocity should also be zero. However, that is not the case, as indicated at the left-hand part of the figure. The reason why force and velocity are "out of phase" is that one of them (velocity) is *cumulative* and must build up, increment by increment, in order to reach its maximum value. The other one (force) is applied by gravity *instantaneously*. (Note that it is dependent on the "resolution of forces," where the strength of the pendulum shaft prevents any gravitational driving force at the very bottom of the swing). This is an easily understood analog of things that happen in electronics, where one thing (not necessarily a "driving force") is instantaneous, and another is cumulative, so their maxima occur at different times.

The times are referred to as the "phases," and they are compared to the time it takes to complete one whole cycle. In 60 Hz (60 cycles per second) ac electricity, one whole cycle takes 1/60 of a second, but another way of looking at it is to consider a cycle as if it was something rotating, going 360 degrees from start to finish. (Actually, the electric generator that makes the ac *is really rotating*.)

In a capacitor, since the voltage is cumulative, it is delayed, and the amount of delay is referred to in terms of degrees instead of by fractions of a second. In this case, the delay (called "lag" in electronics) is 90 degrees. Therefore we can say "the voltage and current are 90° out of phase, and the voltage is *lagging*." Another way to say the same thing is "the current *leads* the voltage by 90°." These relationships are quite important in motors, oscillators, etc.

OPTIONAL
Reactance
Since low frequency ac has difficulty going through a capacitor, there must be something similar to resistance, which the capacitor presents to the electricity. This "something" is called *reactance*. It does decrease the current flow, but it does other things also, and it can be considered to be a sort of overall effectiveness of the capacitor. The "capacitive reactance" (to distinguish it from "inductive reactance," which we will see later) is X_C in equation 9.5.

$$X_C = \frac{1}{2 \pi f C} \qquad (9.5)$$

Pi is the usual 3.14, f is the frequency in Hz, and C is the capacitance in farads.

If an ordinary resistor is put in series with a capacitor, the situation might be as shown in Fig. 9.8. Actually, there will always be *some* resistance (unless superconductors are used!), including the small but often important resistance of the capacitor plates themselves, and various solder joints and contacts.

An optional experiment that the reader might try is to put a 100 ohm resistor in series with the 1,000 mfd capacitor and then do the experiment of Fig. 9.6 over again (but with 150 Ω replacing the 100 Ω shown in the figure). The circular Lissajous figure without the resistor becomes an oval, if 100 Ω is in series. When this resistance is *not* external but instead is inside the capacitor (in the plates, etc.), it can be important in some applications. Even though it might be a "distributed resistance" all throughout the device, one can imagine it to be a single external resistor as in Fig. 9.8. In that case it is called "equivalent series resistance," or ESR. It is part of the specification of high quality devices. The value of the ESR can be calculated from the width of the Lissajous figure (page 98), if precisely calibrated equipment is used. ("Distributed capacitance" can also exist, when two insulated wires are close together, which is often important at high frequencies.)

Another possible case would be resistance that might be in parallel, as shown by the dashed lines in the diagram. It is not practical to measure this with the inexpensive equipment being used in this laboratory course (although the reader is encouraged to try it as an optional exercise), because the 1,000 mfd capacitor offers such a low resistance to 60 Hz ac that only a negligible effect can be seen on the scope when a 100 ohm resistor is put in parallel. (According to eqn. 9.5, this capacitor should have an effect similar to that of a 2.65 ohm resistor.)

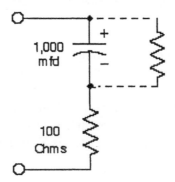

Fig. 9.8 A "lossy" capacitor, with two types of undesirable resistance.

With a series resistance, the total resistance to ac will be more than that of either device alone, and the effective value of this total is called *impedance,* which is Z, as follows:

$$Z = \sqrt{R^2 + X^2} \qquad (9.6)$$

(A rule similar to Ohm's law still applies, so the current still is given by V / Z.) It is convenient in electronics to use a type of vector for this relationship, called a "phasor," as in Fig. 9.8. Note that the Z vector, the long-dash line, is the hypotenuse of the triangle made by the R and X_C vectors (solid lines).

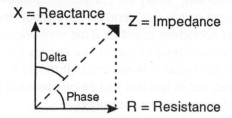

Fig. 9.9 "Phasors," which will be used in Chapter 16 (page 179).

The angle between Z and R is called the "phase angle," symbolized by the Greek letter φ ("phi"), and its value is 45° in Fig. 9.7, but it can be adjusted by changing R. For a pure *capacitance* having zero series resistance, the phase is 90°, but for a pure *resistance* it is zero. The other angle, 90° – φ , is called δ ("delta"). Note that the R / X is the tangent of delta, and in fact it is referred to quite often in discussions about capacitors. Synonyms used in specifications ("specs") include:

R / X = "tanδ "= "loss" = "dissipation factor."

Loss can be calculated from the width and length of the Lissajous figure. Dielectric heaters involve lossy capacitance (see also the middle of page 212).

There is another term that is important in the sharpness of tuning (as with a radio receiver getting all of one station but none of another station). This is the "quality factor," or Q, which is X / R or tan φ.

Still another relationship that the reader might encounter, especially when dealing with electric heaters or motors is the "power factor," which is R / Z. The power obtainable from those devices can be severely limited if the current is too far out of phase with the voltage. There might not be sufficient voltage at the right time to drive high currents through a useful load.

LOSSLESS CONTROL

There are *three* ways to decrease the power going to an electric heater or other device, in order to control it. For a house that is directly heated by electric heating elements mounted in the walls, one way to control the temperature would be to put a giant *rheostat* in series with the heaters. It should be apparent to the reader that such a method would waste a lot of power, right inside the big rheostat, which is itself a heating element (it is "dissipative").

Another way to control the power is to simply turn it *all the way on* when heat is needed and completely off when not needed. This is the way that most home heating (and also air conditioning units) are controlled. However, the variable being controlled (temperature in this case) tends to "overshoot" the desired value. If the turn-on and turn-off operations could be done very often, there would be less overshoot, but the *electromechanical* parts of oil burners and air conditioners would wear out fast if this were done in home units. It should be noted that there is *no resistive loss* of power, and in fact this *fast on-off* operation is actually done in some kinds of *strictly electronic* power controllers.

A third way is to put a big variable-capacitor in series with the heater, because *capacitive ("reactive") limiting of ac has no loss,* at least with nearly-perfect capacitors. This method of controlling ac is hardly ever used, because capacitors that are variable and are "nearly perfect" (practically no ESR) would be very expensive to manufacture. (A similar concept will be described at the end of the next chapter, where a different kind of reactance *is often* used to control power.)

A perfect capacitor does not have the loss expected with a resistor. Therefore, a pure reactance, X, is sometimes treated mathematically in terms of "imaginary" numbers involving the square root of minus one, but we will not use those here.

EQUIPMENT NOTES

Components For This Chapter	Radio Shack Catalog Number
Battery, nine volts	23-653
Clip leads, 14 inch, one set.	278-1156C
Multimeter (or "multitester")	22-218
Capacitor, 1000 mfd	272-1019 or 272-1032
Potentiometer, 5K	271-1714 or 271-1720
Resistor, 100 ohm, 10 or 1 watt	271-135 or 271-152
Resistor, 150 ohm (for optional exper., page 100)	271-306 or 271-308
Oscilloscope, Model OS-5020, made by LG Precision Co., Ltd., of Seoul, Korea	

CHAPTER 10

Inductors

ELECTROMAGNETS AND GENERATORS

The reader probably remembers from previous courses in general science that an electric current going through a wire will produce magnetism. The "magnetic field" is around the wire, somewhat like an invisible tube. This is shown diagrammatically by the two circles at the upper left of Fig. 10.1. When the wire is coiled, as at the right-hand side of the figure, the fields line up to make a

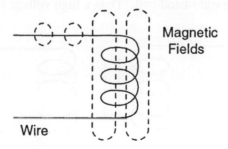

Fig. 10.1 Magnetic fields (dashed lines) around current-carrying wire.

"donut" or "toroid" of magnetic fields, much stronger than the fields around the single wire portions.

If an iron rod (not shown in the diagram) is poked vertically down through the coil, the fields can go through that iron much easier than through air or vacuum.

The ease of producing a strong field is called "permeability," and it is somewhat similar to the "dielectric constant" described on page 92. Iron is easily "polarized" magnetically, because it has many "unpaired" electrons that are spinning. An electron spinning generates a small magnetic field of its own. The spins can be lined up in the same direction, if they are put in a field such as the one from the wire. In some ways this is similar to the charges lining up inside the ceramic "dielectric" shown on page 91, thus raising the capacitance.

Of course, an iron core plus a coil of wire can be an "electromagnet." When the current is first turned on, the field suddenly expands, from zero to whatever the final size will be at the maximum current.

The reader probably remembers, also, that if a second wire is moved through a stationary magnetic field, *or if* the field is moved across a stationary wire, either way, a new voltage will be generated in that second wire. This is the principle of the "electric generator," which is used to make most of the electricity that comes to our houses and factories. A convenient concept is that "lines of magnetic field are being cut by the wire." (The field really varies smoothly with distance, and it is not grainy like wood. But for convenience we imagine lines, where the more lines, the stronger the field.) A *higher voltage* will be generated if *more* of these conceptual "lines" are being *cut per second.*

If there are *two* coils, *both stationary*, as in Fig. 10.2, and an electric current is *suddenly turned on* in the left-hand coil, the magnetic field lines will quickly expand from zero to the final size, and they will get "cut" very fast by the stationary wires of the right-hand coil. Thus a high voltage will be generated in that coil.

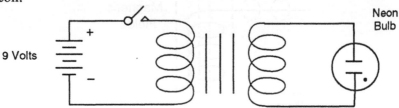

Fig. 10.2 A transformer, with momentary dc in the primary coil.

If the current in the left-hand ("primary") coil is *turned off suddenly*, the field lines will be cut as they collapse back to zero size, but in the opposite direction. The voltage in the right-hand "secondary" coil will be generated with opposite polarity. (It could be a high voltage, if the turn-off is fast enough.) If alternating current is put through the primary, then ac voltage will appear in the secondary, and this of course is a typical use of the "transformer."

If there are *more turns in the secondary* coil, each turn is a little voltage generator in series with the others, so a *higher voltage* will appear across the terminals of the secondary. However, the amount of *power* coming out must be the *same* as what goes in, because power is energy per time (see the bottom of page 16), and energy can not be made or destroyed — it can only be changed in its *form*. Since power is voltage times current (eqn. 2.8 on page 17), if there is a *higher voltage* in the secondary than in the primary, then there must be a *lower current,* in order for the energy per second to be constant. These are in direct proportion to the ratio of turns, primary versus secondary.

On the other hand, if there are *fewer turns in the secondary* than in the primary, then there will be a *lower voltage*. However, *more current* could then be gotten from the secondary, again depending on the ratio of turns. This is analogous to a mechanical *lever,* where higher speed can only be at lower force.

EXPERIMENTS

Transformers

Ordinarily, the *primary* coil of the transformer used in this course would be plugged into a 120 V ac wall socket, and only 12 volts of ac would be available in the secondary, because that has ten times fewer turns than the primary has. Thus, this transformer is usually in a "step-down" mode, and in fact it was used that way in the optional experiment on page 98 of the previous chapter.

If the circuit of Fig. 10.2 on the previous page is assembled with the 9 volts going into the *primary* (two heavily insulated black wires), it will still be used in the "step-down" mode, and the neon bulb *will not* light when hooked up to the secondary (lightly insulated yellow wires), because it requires much more than 9 volts, and instead it will probably be getting less. (Actually, it *could* be lit if the battery was connected or disconnected *fast* enough to make up for the unfavorable ratio of turns, but probably neither operation can be done fast enough here.)

The neon bulb *will* flash brightly if the *battery* is connected to the original *secondary* (yellow-insulated wires of the transformer) and then *suddenly* disconnected, and the *bulb* is attached to the original *primary* (thicker wires with black insulation), because everything is favorable for a high voltage output— the *step-up,* and a *fast* current decrease in the primary. There will be less current out, but it only takes a very low current to operate the neon bulb, as was seen on page 6 of Chapter 1. (Note that the ignition coil in an automobile works this way, but with far more turns in the secondary. With not so many turns, it is used in "switching power supplies," including "inverters" that change dc to ac.)

Autotransformers and Inductors

Figure 10.3 shows only the 12 volt side of the transformer being used, and the 9 V dc is going to one end (yellow) plus the "center tap" (thin black insulation). The experimenter should attach the oscilloscope with the *vertical* being *voltage* (CH2/Y), versus the *horizontal* being *time* (TIME/DIV = 0.1 s) hookup. The input MODE is CH2, the trigger MODE is AUTO, and the TRIGGER SOURCE is VERT (similar to pages 76 and 77). The VOLTS/DIV can start out at about 1 but be changed as needed, including rotating the central VARIABLE knob.

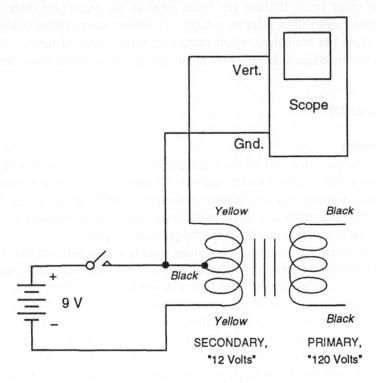

Fig. 10.3 An "autotransformer" (which is type of inductor). Note again that two wires crossing *without* a large black dot indicates *no* connection.

As usual in this course, there is no real switch, and a clip lead is simply touched for a short time ("momentarily") to a wire or terminal, to put a short burst of current through half of the "12 volt" coil (ordinarily the secondary, but acting as a primary here). When the current is suddenly increased, a *downward* spike appears on the scope, showing that the "top" half of the center-tapped coil has *negative* voltage going to the scope hook, and positive going to its grounded clip lead. This is just the opposite of the voltages on the "bottom" half of the coil. Therefore the two halves tend to cancel each other, whenever we try to suddenly increase the current, which tends to *prevent the current from increasing quickly.*

The next step is to suddenly "open" the "switch," and an *upward* spike appears, meaning that a voltage has been generated in the top half-coil that *adds* to the bottom half's voltage. The overall action therefore tends to *prevent the current from stopping quickly.*

COMMENT

Several things can go wrong with this experiment, although it should be possible to make it produce the exact result described above. First of all, the experimenter might verify that an *"upward"* spike on the scope display does indeed mean a *positive* voltage on the BNC cable's *hook.* (Be sure that the NORM / INV button has not been pushed accidentally.) This can be checked by attaching the hook and ground clip directly to the battery, with nothing else attached, and seeing whether the bright spot in the display goes upward. (In fact, the experimenter should get in the habit of "checking out" the scope like this, whenever results do not match expectations.) Secondly, sometimes a capacitance inside the scope gets charged up and then will not respond for a second trial. If that seems to be the case, discharge it before each new trial, by momentarily touching the hook to the ground clip. Thirdly, sometimes safety devices inside the scope prevent spikes of certain polarities. If the downward spike appears but the upward one will not, try reversing the polarity of the battery, and also pushing the NORM / INV button. Then both spikes should become visible, in this doubly inverted mode. Another point is that very little current flows into the scope, because of its extremely high "input impedance." If a low resistance such as an ammeter is attached to the top half-coil, then considerable current will flow, and this will make its own magnetic field, which interacts with the field of the bottom half-coil in a complex manner. The voltages of the two halves might then become additive and not cancel.

The action of the top and bottom halves of the coil is similar to the action of the primary and secondary of an ordinary transformer, except that in the above experiment they are parts of the same coil, and they have the same numbers of turns. In some cases it is desirable to use just one continuous coil with a "tap," but not necessarily at the center, and in fact it can be movable, to provide varying voltages. This is called an "autotransformer," and it is often used to convert 120 volts to either lower or higher voltages, all ac, for operating small heaters or stirring motors in chemical laboratories. The brand name of one such commonly used device is Variac®.

LESSON

The main idea of this experiment is that a fast-changing current in a coil will "induce" an opposite voltage in a nearby coil, and *also in itself.* The latter effect is what causes an inductor to resist change in the current going through it, which produces the electrical analog of mechanical inertia, including the "inductive kick" seen in Chapter 1. It does not have to be center-tapped at all, and in fact it usually is not. When the current changes fast, every turn of a coil induces an opposing voltage in each other turn, and this causes "inductance." By comparison, direct current goes through an inductor with no more resistance than it would have in a straight wire with no metal core — in other words, easily. But the higher the frequency (faster change in direction), the less current gets through an inductive coil, especially if it has an iron core to enhance the magnetic action.

A mechanical analog of an inductor is shown in Fig. 10.4. As briefly mentioned on the very first page of Chapter 1, an inductor acts like a heavy piston inside a water pipe, tending to prevent fast changes in flow, either increases or decreases. There is very little resistance to steady water flow, similar to dc electricity in an inductor. On the other hand, high speed changes in direction would meet great resistance, similar to high frequency ac in the same inductor.

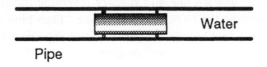

Pipe

Fig. 10.4 A piston analog of an inductor, with an O-ring seal at each end.

Shorted Additional Coil

The experimenter should construct the circuit of Fig. 10.5 on the next page. At first, the wires shown as dashed lines are not connected. The experiment is done as on page 7 of the first chapter, where the "switch" is opened suddenly, and there is a bright flash in the neon bulb. Then the short circuiting clip lead (dashed line in the diagram) is attached to the transformer's 120 volt coil (heavily insulated black wires or soldered-on power cord). Now when the switch is opened, there is no neon flash. What is happening is that current is induced in the "shorted" coil, as the battery current suddenly decreases. This new current has a magnetic field that tends to *stop the current faster,* whenever the battery current decreases. That of course is the opposite of tending to keep it going, as the coil turns on the other side of the transformer do to each other.

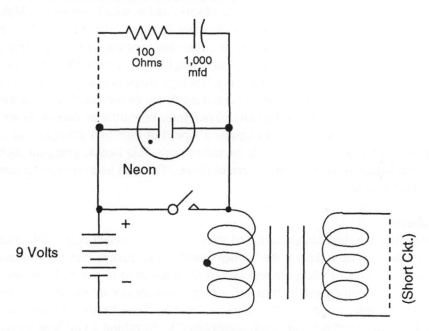

Fig. 10.5 The inductive kick experiment, with a few new items.

There is a very important consequence of this behavior. First of all, if there is *no short circuit,* then the left side has a high inductance, which resists the flow of ac current. (Of course, we are not using true 60 Hz ac here, but transformers usually do.) Therefore, when there is no output (no "short" on the right-hand side, as shown here), very little current flows into the input, and the system is just "idling." A transformer can be plugged into a source of ac and hardly draw any power under such conditions, while it is waiting to be used. When a "load" (such as an electric motor) is later attached to the output side, current flows through both sides, and of course power is then being drawn from the source, possibly a wall socket. It is only drawn from the input side as it is used at the output side, so it does not waste power if it is not being used.

If there is an extreme case such as a *short circuit* in the output, as we intentionally produced here in order to make the effects easy to observe, then there will also be a short circuit at the input. This could burn out either side.

Something similar happens with an ac electric motor, which can spin at high efficiency if it is not being used, but which draws far more current if it is slowed down by heavy use. It will become a short circuit and burn out, just like the input side here, if an excessive mechanical load completely prevents rotation. The reason, briefly, is that a motor rotating fast has high frequency ac going through its inductors, and the inductors (various coils) inside the motor tend to prevent the current from changing direction so fast. Therefore, very little ac current flows. When the motor slows down, the frequency (speed of direction change) slows, and more current can get through. If the motor stops, it is like dc going through its internal inductors, and too much current flows. This will be explained further in the chapter on motors.

Snubbers

Disconnecting the short circuit wire, and then opening the switch again will make a spark and also a neon flash at the "switch" itself. This spark is very hot and can do a lot of damage to whatever real switch (as contrasted to our make-believe "switch") is actually used to turn off a large transformer or motor or other inductor. In order to prevent the spark, a "snubber" can be used (sometimes called a "spark suppressor" or "spike suppressor"). Attaching a clip lead where the upper-left dashed line is shown in Fig. 10.5, and thus putting the capacitor and resistor across the "switch," the capacitor will absorb the inductive kick and prevent the high voltage spark. The inductive current will still flow for a short time after the switch is opened suddenly, but it goes into the capacitor instead of making a spark. There is no longer enough high voltage to light the neon bulb, which further proves that the snubber is doing its job.

The 100 ohms is high enough to prevent excessive initial current from damaging the capacitor, but low enough to soak up the inductive kick. If the capacitor is used many times in a row, it will eventually charge up and then fail to work for snubbing the sparks. In that case, a 1,000 ohm resistor should be put *across* the capacitor to slowly discharge it. (In most real equipment, the capacitors are only about 100 microfarads (mfd), and they have a resistor for discharging.) There are other types of spark snubbers, which will be explained in the chapter on diodes.

Symbols

Another way to draw coils is shown in Fig. 10.6. It is often used in modern publications, because it is easier to draw than complete spirals.

The two black dots on the left-hand transformer indicate that, at a certain time, the current is coming "upwards," out of *both* sides. Since alternating current goes both "up" (as shown here), and also "down" a fraction of a second later, it is pointless to say "+" or "-" and, instead, we describe the left-hand transformer as "being in phase." In the right-hand transformer, the two coils are wound differently from each other, so at any time when "positive" current is going "up" in the primary coil, it is going "down" in the secondary. We describe this as "being 180 degrees out of phase." (The reader might refer to Fig. 10.8 for further clarification.) The black dots are only shown in those cases where this is an important item.

In Phase   180 Degrees Out of Phase

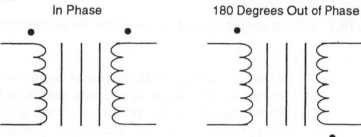

Fig. 10.6 Coils drawn a different way. (Also, the dots show phases.)

Voltage Versus Current

While the oscilloscope is still hooked up, the experimenter can change the circuitry to that of Fig. 10.7 (on the next page), which is the same as on page 97, except for the transformer primary being substituted instead of the capacitor. The 120 volt coil of the transformer can be used here, either with or without its long power cord. On the scope screen, the *voltage* (vertical) will be observed to increase *before* the current (horizontal) increases. The VOLTS/DIV knobs may have to be set for more sensitivity (lower voltages per display division), because even the multi-turn 120 volt winding will not have as much inductive effect as the 1,000 mfd capacitor had capacitance effect. In this case, the *voltage* can be applied to the coil immediately, but the *current* can not increase so fast, just like water pressure on a heavy piston can not get it up to full speed instantly. An increase in current induces a high reverse voltage, so the initial current increase is retarded in its timing.

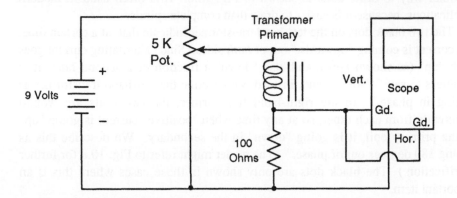

Fig. 10.7 Voltage versus current "curve tracer" for an inductor.

Phase

As observed in the curve tracer experiment, the current lags the voltage, and the voltage is said to lead the current, when an ac sine wave goes through an inductor. (Doing the experiment requires two transformers, so it will not be performed here.) The relationship is shown in Fig. 10.8. Actually, the pendulum on page 98

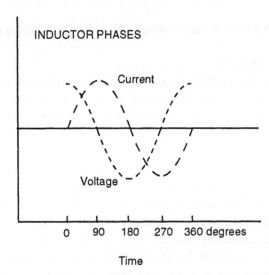

Fig. 10.8 Sine wave ac in an inductor, showing lead and lag.

is more closely analogous to the inductors in this chapter than to the capacitors in the previous chapter, but the general ideas of instantaneous ("leading") versus cumulative ("lagging") behavior apply to both.

OPTIONAL
Current Versus Time
When the voltage is first applied, and the current builds up, that current follows the following equation:

$$I = I_{max} (1 - e^{-t/\tau}) \tag{10.1}$$

where t is time and τ is the time constant, "tau" (the time when the current gets to about 2/3 of its maximum value). This is similar to the capacitor example on page 94, but with current changing now instead of voltage. The curves also are similar, but with *inductor current* substituted for *capacitor voltage.* The voltage itself is given by Ohm's law, $V = IR$. (For "e," see page 95.)

When the driving voltage is removed, the current decays according to:

$$I = I_{max} (e^{-t/\tau}) \tag{10.2}$$

where τ is the time at which the current gets to about 1/3 of I_{max}.

Although $\tau = RC$ with capacitors (page 95), $\tau = L/R$ with inductors, where L is the inductance. Thus a resistance tends to decrease the time constant with inductors, while it increases it with capacitors.

The *inductive kick,* that is, the voltage generated by decreasing the current quickly, is $V = L[$(current change) / (time change)$]$. Thus, the faster the change in current, the higher the voltage that is generated, almost without limit.

Reactance
There is an inductive reactance, similar to the capacitive reactance described on page 99. The formula is

$$X_L = 2\pi fL \tag{10.3}$$

where f is the frequency, and L is the inductance.

The unit of frequency, f, is the Hertz, named after the early scientist, Heinrich Hertz, and the unit of inductance, L, is the Henry, named after the early scientist, Joseph Henry. The vector diagram on page 101 can still be used to show the relationship between inductive reactance, X_L, and resistance, R, with the hypotenuse being the impedance, Z. Similarly, the "loss" is the tangent of the angle, delta. The power factor for inductors is also R / Z, as on page 101.

Inductance

The value of the inductance, in units of microhenrys, can be calculated approximately from the dimensions of the coil shown in Fig. 10.9, as follows:

$$L = \frac{\mu \ (\text{turns})^2 \ (\text{radius})^2}{9 \ (\text{radius}) + 10 \ (\text{length})} \qquad (10.4)$$

where the radius and length are in inches. The Greek letter μ ("mu," pronounced "myoo") is the relative permeability, compared to the permeability of vacuum. This would be essentially the μ of the iron (or "ferrite" ceramic) core. (The μ of air is close to that of vacuum and is ignored in air-core coils. See also page 268.) The symbol at the left, the circle with a sine wave inside, stands for any source of

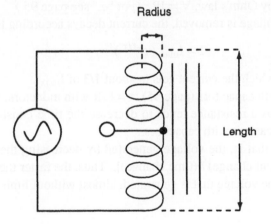

Fig. 10.9 An inductor.

alternating voltage or current, such as the output of a transformer or generator or oscillator. (The latter will be covered in a later chapter.) The center-tap has no effect on the inductance, if it is not attached to anything.

COMMENT

Inductors tend to be bulky and heavy, and in many cases they can be replaced by smaller, lighter weight capacitors (as will be described later, on page 123). That is why, *instead of inductors,* circuit designers try to use capacitors or "gyrators" (to be mentioned briefly on page 125).

LOSSLESS CONTROL

Inductors decrease ac current by "reactance," similar to the way capacitors do (see last text paragraph on page 102). If the inductor were perfect, and had no resistance, then no power would be lost through heating ("dissipation"). Nearly perfect inductors are nearly lossless. For this reason, the autotransformer described at the bottom of page 107 can decrease ac voltage and current with very little wasted power. Also, a simple two-terminal "ballast" inductor, which is not center-tapped, is often used to limit the current in fluorescent lights.

Another way to do this, as also mentioned on page 102, is by fast on-off switching. This is done with SCR controllers, as will be described Chapter 21.

If a large rheostat is used, however, a great deal of power is wasted in the form of heat dissipation. That method was used in old electrical equipment such as early electric railroad engines and trolley cars, before it was realized that inductors could be used and even earlier, when most electric power was dc.

SATURABLE CORES AND MAG-AMPS

As the alternating current input of a transformer increases, the output also increases, but it reaches a maximum value when the core becomes "saturated" and can no longer increase its magnetization. If the transformer is purposely operated under these conditions, the output will be constant, even if the input varies a small amount. This is sometimes used to provide a dependable, constant voltage to operate sensitive instruments, heaters, and other devices, and it is called a "saturable core reactor" or a "constant voltage transformer." In newer equipment designs, it has mostly been replaced by the semiconductor devices that will be described in Chapter 15 (Fig. 15.6).

If a transformer is operated a little bit below saturation, and a small *direct current* is sent to a *third* coil, this can be used to sharply change the output, by pushing the total magnetization up into saturation. (None of the dc itself becomes part of the output — it only *affects* the ac, but it is not part of it.) Then, by decreasing the dc a small amount, the iron core can become *unsaturated,* and far more ac will go through, into the output. Thus a small dc signal can control a large ac "throughput." This is called a "magnetic amplifier," or "mag-amp." It is used to control high currents for heaters in older industrial furnaces, although SCRs are used instead, for modern designs. The SCR and similar devices will be described in Chapter 21.

EQUIPMENT NOTES

Components For This Chapter	Radio Shack Catalog Number
Battery, nine volts	23-653
Clip leads, 14 inch, one set.	278-1156C
Transformer, 120V to 12V, 450 ma, center tapped	273-1365A
Resistor, 100 ohm, 10 or 1 watt	271-135 or 271-152A
Potentiometer, 5K	271-1714 or 271-1720
Capacitor, 1000 mfd	272-1019 or 272-1032
Neon test light, 90 volts	22-102 or 272-707
Oscilloscope, Model OS-5020, LG Precision Co., Ltd., of Seoul, Korea	

CHAPTER 11

Filters and Resonance

SIMPLE RC FILTERS

A filter allows ac of some frequencies to pass through it with the amplitude (voltage or current) essentially unchanged, but the amplitudes of other frequencies are *decreased*. (An "active filter," involving transistors, can *increase* certain frequencies, as well as decreasing some.) A simple example, using just one resistor and one capacitor, which is called a "low-pass RC" filter, is shown in Fig. 11.1. It is a type of "first order Butterworth" filter.

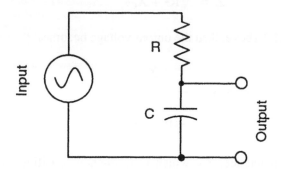

Fig. 11.1   A first order Butterworth low-pass RC filter.

As mentioned previously (pages 56 and 114), the circle symbol represents a source of ac voltage. The resistor and capacitor together make a voltage divider, similar to a potentiometer. However, the capacitor part of it resists the flow of current differently, varying according to the frequency of the ac. At *higher* frequencies, as discussed in Chapter 9, it allows more ac to go through it. In this diagram, the "output" voltage is therefore *lower*.

At *low* frequency, the capacitor acts as a very high resistance, and with dc it is an "open circuit" (essentially infinite resistance). Therefore, *more* voltage will appear across the output terminals — hence the name "low-pass."

On page 99 of the Capacitors chapter, the concept of *reactance* was explained. This is a sort of *effective resistance* for a capacitor in an ac circuit, and its symbol is X_C. Since the RC filter is similar to a potentiometer with a pair of resistors (page 43), the reader might expect the output voltage, as shown in the diagram on the previous page, to be the input voltage times X_C, divided by the sum of R plus X_C. However, resistance does *not* add to reactance directly. Instead, we use a sort of *effective total resistance,* which is the impedance, Z. Thus the output voltage is given by:

$$V_{out} = V_{in} \frac{X_C}{Z} \tag{11.1}$$

The impedance, Z, was given by equation 9.6 on page 101, as follows:

$$Z = \sqrt{R^2 + X_C^2}$$

Putting Z into eqn. 11.1 above, then the output voltage becomes:

$$V_{out} = V_{in} \frac{X_C}{\sqrt{R^2 + X_C^2}} \tag{11.2}$$

The capacitive reactance, X_C, which is frequency sensitive, was given in equation 9.5 on page 99, and it is $X_C = 1/(2\pi fC)$. This can be "plugged in" to the above equation at any frequency, to see how the filter will affect the voltages.

OPTIONAL

The 2π comes from the fact that the mathematics of sine waves are closely related to the behavior of a vector rotating around one of its ends, like the hour hand of a clock. The vector diagram on page 101 is an example of this relationship. The 2π factor is the number of radians in a complete circle. Instead of f, electrical engineers often use the "radial frequency," ω (the Greek letter omega), where ω is the number of rotations of the vector per second. (This is sometimes called the "angular frequency.") Also, $\omega = 2\pi f$, so ωC can be used instead of $2\pi fC$, which simplifies many complex formulas.

There is some frequency at which the "effective resistance" of the capacitor (the *reactance*, X_C) equals the resistance. That is a convenient reference point for the study of filters. It is sometimes called the "corner frequency." (Some other synonyms are listed on page 122.)

For a one volt ac input to the simple filter of Fig. 11.1, the output is shown in Fig. 11.2 . The "transfer function" is the output divided by the input, for some particular frequency. As shown in the figure, at a frequency of "1," it is 0.707. (Sometimes it is called the "gain," even if it is less than one.)

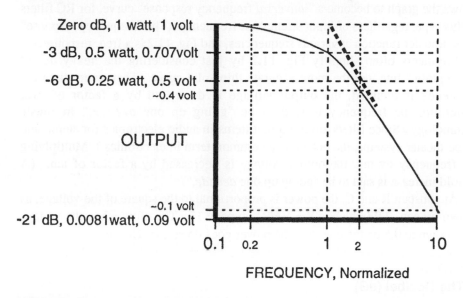

Fig. 11.2 labels:

Zero dB, 1 watt, 1 volt

-3 dB, 0.5 watt, 0.707volt

-6 dB, 0.25 watt, 0.5 volt

~0.4 volt

OUTPUT

~0.1 volt

-21 dB, 0.0081watt, 0.09 volt

0.1   0.2   1   2   10

FREQUENCY, Normalized

Fig. 11.2 Output of the filter in Fig. 11.1, when $R = X_C$. Both the transfer function and the frequency are normalized in this diagram. The frequency labeled "1" is really $1/2\pi RC$. (Log V versus log f is called a "Bode plot.")

In this type of graph, both axes show "normalized" values. That is, the transfer function is shown for an input of "one," and the frequency is centered around a value of "one." If the input in some experiment with this kind of filter is

3 volts instead of 1, then all the values on the vertical axis are multiplied by 3, and the output for a frequency of "one" is $3 \times 0.707 = 2.121$ volts. If the input is 3 watts, then the output is $3 \times 0.5 = 1.5$ watts at that frequency.

But what is the frequency that is indicated by the mysterious "1" value? It depends on R and C. This diagram is only correct for cases where R of the resistor and the "effective resistance" (the reactance, X_C) are *equal*. (However, it could easily be adjusted for unequal values, by using eqn. 11.1.) In that particular case, $f = 1/2 \, \pi RC$ (as could be derived from eqn. 9.5 on page 99, by setting R equal to X_C, and then having R and f switch places.).

Suppose R and C have values such that $f = 300$ Hz. Then the "1" on the horizontal axis of Fig. 11.2 corresponds to 300 Hz, and the "2" represents 600 Hz, the "0.2" represents 60 Hz, and so on.

The word "normalization" means dividing the true values by some number so that an important position on the graph becomes 1. In this case it has been done for both the middle of the horizontal axis and the top of the vertical axis. That allows the graph to become a "*universal* frequency response" curve, for RC filters of this type, regardless of input strength or frequency. (The "frequency response" is the transfer function at various frequencies, and Fig. 11.2 is a freq. response.)

Engineers often simplify Fig. 11.2 by just considering the heavy dotted "asymptotic" line, not the more accurate thin solid curve. Then every time the frequency is doubled, the output voltage is decreased by a factor of two. (Doubling the frequency is said to be "going up one *octave*," in music terminology. Since filters are important items in audio electronics for stereo and home theater systems, the octave is a common term in electronics.) Multiplying the frequency by ten, the output voltage is decreased by a factor of ten. (A tenfold increase is said to be "going up one *decade*.")

At constant R and C, the power is proportional to the square of the voltage, as shown by eqn. 2.9 (which is $P = V^2/R$) on page 17. Therefore when the voltage goes down to 0.5 on the diagram, the power goes down to 0.25 watt.

OPTIONAL
The Decibel (dB)
At the left edge of Fig. 11.2, the output values are expressed in units of dB, or "decibels." This is a term commonly used in electronics as applied to sound reproduction and also in fiber optics. As mentioned on page 44, animal sensors such as ears, eyes, and taste buds respond to an enormous range of intensities, more than 10^{11}, and the only way this could be accurate at the bottom end of the intensity range is if the response is logarithmic. (Logs greatly expand the bottom end of the scale, separating the weak signals from each other.)

Natural evolution has provided animals with logarithmic sensors, and therefore electronic technologists use a log scale to express differences in sound loudness, and also in the corresponding electrical signals. The ear responds to power per eardrum area, and the latter can be kept constant in a given set of experiments. A convenient logarithmic ratio of power differences is the "bel," named after Alexander Graham Bell, the inventor of the telephone (and founder of AT&T Corp., the author's employer for many years.)

One bel is the $\log_{10}$ of the output wattage divided by the input wattage. If the output is 10 W and input is 1 W, then there is 1 bel of "gain" (equal to the "transfer function"). If the output is 100 W with the same 1 W input, then the gain is 2 bels. If the output is only 0.001 W, then the gain is –3 bels.

To the human ear, 2 bels sounds about twice as loud as 1 bel, so it takes about ten times as much power to sound roughly twice as loud. However, the ear can distinguish sounds that differ by only about 0.1 of a bel, so engineers use units of a tenth of a bel, or a "decibel," abbreviated "dB." (The B is capitalized, because it is the first letter of someone's name, although this is not always done with other terms such as amperes, as in ma for milliamps, or in bels when used alone.)

The number of dB is ten times the number of bels, so

$$dB = 10\log_{10}(W_{out}/W_{in}). \tag{11.3}$$

However, power is proportional to the *square* of *voltage* (or *current*), so

$$dB = 20\log_{10}(V_{out}/V_{in}), \text{ or} \tag{11.4}$$

$$dB = 20\log_{10}(I_{out}/I_{in}). \tag{11.5}$$

Some of the decibel ratios are summarized in the following table:

dB	Power Ratio	V or I Ratio
0	1 : 1	1 : 1
1	1.25 : 1	1.12 : 1
3	2 : 1	1.41 : 1
6	4 : 1	2 : 1
10	10 : 1	3.16 : 1
12	16 : 1	4 : 1
20	100 : 1	10 : 1

The expression "3 dB down" means -3dB, or half the power. The unit dBm is the output compared to 1 milliwatt input.

A few commonly used relationships that appear in Fig. 11.2, using the simplifications of the heavy dotted line, are as follows:

Increasing the frequency by an octave, halves the output voltage;

Increasing the frequency by a decade, lowers the output voltage by a factor of ten.

However, these are quite inaccurate until higher frequencies are reached, well above the "*corner* frequency" mentioned on page 119. That frequency is usually considered to be the thing marked "1" in the diagram, although the reader could probably argue that it ought to be the one marked "0.1." At any rate, it is where the curve begins to get fairly steep. Other terms that all mean the "1" frequency are "*half-power* point," and "*turnover* frequency," and "*roll-off* frequency," and "*cut-off* frequency," and particularly in England, "*dividing* frequency."

EXPERIMENTS

Low-Pass RC
Since this course does not use a source of varying frequencies, a very simple experiment will be done with whatever electrical noise is "ambient" (generally present in the immediate environment).

The oscilloscope is set up as on pages 76 and 77. Put a clip lead on the hook terminal and another on the ground clip. The VOLTS/DIV can be made more sensitive (about 50 mV), until a vertical signal appears, and the TIME/DIV (about 2 ms) can be adjusted by means of its VARIABLE until a single sine wave appears. This is noise picked up from the surroundings, mostly 60 Hz ac (in the U.S.A.), with various other distorting harmonics added to it. Some very high frequency noise might also be visible (closely spaced peaks). Now attach a 100K resistor and a 1,000 mfd capacitor in series, with the scope hook (vertical input) attached to the junction between them. Clip the scope ground onto the other end of the capacitor, so the ckt. looks like Fig. 11.1, with the two long clip leads functioning as the ac source, and the scope attached to the filter's output. The signal on the scope should be very much decreased in amplitude. The corner freq. is $1/2\pi RC$ (top of page 120), which is $1/(6.28 \times 100K \times 0.001) = 0.0016$ Hz. The frequency of the 60 Hz noise is so much higher than this, that the "attenuation" (decrease in voltage) will be quite close to 100%. (A 4.7 mfd capacitor plus a 5K pot set to about 560 ohms would have a 60 Hz corner frequency. Or, with a 0.1 mfd capacitor and 1K, the 60 Hz noise would go through the low pass filter.)

High-Pass RC

Switching places between the 100K resistor and the 1,000 mfd capacitor makes it a high-pass filter, as shown in Fig. 11.3(B), below. The normalized output curve would look like Fig. 11.2, if reflected side-for-side by a mirror, so the low part is on the left at a normalized frequency of 0.1, and the flat, high part is on the right, at a frequency of 10. The 60 Hz noise from the surroundings will go through this filter and be visible on the oscilloscope. This experiment should be carried out.

NOTE

Any time the scope trace disappears, turn the VOLTS / DIV towards the left, to 5 V, and switch the TRIGGER SOURCE to LINE. Adjust POSITION, Y slowly up or down all the way, until the visible line appears again. Then adjust these knobs back toward the desired settings, in small steps.

OTHER FILTERS *(No experiments until page 128)*

RL Butterworth Filters

An inductor has an effect that is opposite to the capacitor's effect. This is shown in Fig. 11.3 (C and D), for direct comparison with their capacitor counterparts.

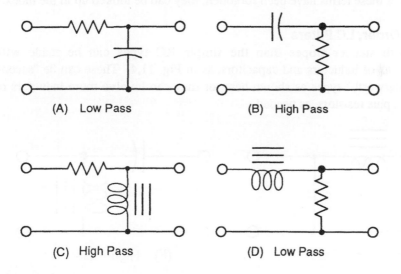

Fig. 11.3 RC and RL filters compared.

The mathematical relationships are similar to those of a capacitor, but the inductor's effective resistance ("reactance") is given by eqn. 10.3 on page 113.

In this course, a second inductor other than the transformer would be needed to do experiments with inductive filters. That is not part of this simplified course, so such experiments will not be done, with the possible exception of an optional one in the diodes chapter.

One type of very simple inductive low-pass filter is becoming common in electronics. This is a small magnetizable ring made out of "ferrite," a ceramic material similar to the natural rock called "lodestone" or "magnetite." Magnetite is Fe_3O_4, and it is difficult to magnetize or demagnetize. (Its magnetic characteristics are similar to those of the *hardened* iron in a knife blade.) Ferrite has some of the iron replaced by zinc and either manganese or nickel, which makes it very easy to magnetize, and also it spontaneously demagnetizes when an external field goes to zero. (This is called a *"soft"* characteristic, similar to that of the soft iron core in a transformer. In fact, ferrite is often used in transformers.)

The soft ferrite rings, called "beads," are often placed around wires that go in and out of computers. The low-pass filters that result are just strong enough to keep EMI from getting into the computer via its wires, and also to keep unintentional RF from getting out and interfering with TV sets, etc. (If the meanings of these terms have been forgotten, they can be looked up in the index.)

Second Order, LC Filters

Filters with steeper slopes than the simple RC types can be made with *combinations* of inductors and capacitors, as in Fig. 11.4. These can be "second order Butterworth" types as shown here, or they can involve more inductors or capacitors, plus resistors, as needed.

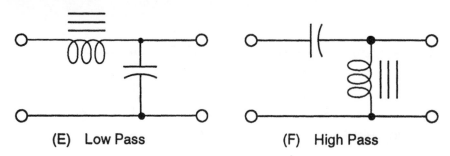

    (E)   Low Pass   (F)   High Pass

Fig. 11.4   Second order LC filters.

OPTIONAL

The impedance of a circuit containing R, C, and also L is given by:

$$Z = \sqrt{R^2 + (X_L - X_C)^2} \tag{11.6}$$

Note that the effect of the capacitor is negative, because it is "out of phase" with the inductor.

A set of filters of adjustable effectiveness, each tuned to a different octave, is familiar to many readers, in the form of the "graphic equalizer." This is a component that can be included in an audio music reproduction system, to adjust the "frequency response" of the whole system. It was originally used mainly in recording studios, but in 1976, the first full-length article in a popular magazine led to its becoming a commonplace item.[1] Instead of inductors, modern equalizer designs tend to use transistor circuits called "gyrators," which are lighter weight and offer more flexible adjustment possibilities.

A consequence of the availability of high quality equalizers is that *most* inexpensive stereo systems that had previously-mysterious sound qualities such as "glassiness" or "harshness" can usually be made to sound as good as *most* very expensive systems. An inexpensive graphic equalizer is simply used to make both have the same frequency response (much to the dismay of some manufacturers of expensive systems). This claim can only be proved with the use of an "equalized double-blind test," which was innovated by the author and reported in popular magazines[2] in 1980. The use of this test has now spread worldwide among stereo circuit designers, something like a computer virus, and it has taken on a "life of its own." (Possibly that is an unpleasant metaphor, but it is an accurate comparison.) A word which has suddenly become popular among sociologists is "meme," which means[3] any new idea or behavior pattern that spreads on its own, again being "like a computer virus," and it can be either a good idea or a bad one. The equalized double-blind test, making use of inexpensive but good quality filters, is such a meme.

The simpler types of filters have phase lead or lag that varies with frequency, and in some applications this is undesirable, examples being the crossovers in loudspeakers. Various complex filters are designed to minimize this effect.

---

[1] Dan Shanefield, "Audio Equalizers," Stereo Review, May 1976, p. 64. Also, May 1996, p. 112.

[2a] Dan Shanefield, "The Great Ego Crunchers: Equalized Double-Blind Tests," High Fidelity, March 1980, p. 57 (available in most public libraries on microfilm). [2b] S. P. Lipshitz, et al., Jrnl. of Audio Engineering Society, Vol. 29 (1981) p. 482, particularly ref. 27 (in university libraries).

[3] Susan Blackmore, "The Meme Machine" (see N. Y. Times Book Review, April 25, 1999, p. 12).

Another problem with filters, besides phase changes, is "ringing." Once an ac signal goes into a filter, it tends to continue electrically vibrating ("oscillating") to some degree, even after it is supposed to have stopped. A mechanical analog is the vibration of piano strings, after they had been struck. (The stretchiness of the string is the mechanical analog of an electronic capacitor, and its inertia is the analog of an electronic inductor.) Soft "snubbers" or "dampers" touch the piano strings to suppress this ringing, unless a foot pedal is pushed, to remove the snubber and purposely allow continuation ("sustaining") of the musical tone. In electronics, resistors are sometimes added to the circuit to suppress ("dampen") the ringing and minimize it, and other methods are also used.

RESONANCE

There are many situations were ringing is desirable in electronics. The tendency to continue oscillating is called "resonance." As mentioned above, it is analogous

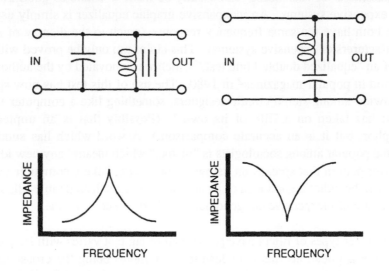

Fig. 11.5 Two simple resonant circuits.

to a mechanically stretchable spring, plus a mechanically inertial mass. Combining the spring and lightweight piston in the water pipe shown on page 89 with the heavier one on page 108, a vibrating system is likely to result, at least to some degree.

Two simple types of electrical resonance are series and parallel, as shown in Fig. 11.5 on the previous page. In the *series* case, when the voltage of the *inductor* is at a *maximum* (at the beginning of the cycle on page 112), the voltage of the *capacitor* is at *zero* (page 98), so the inductor's voltage drives current right through the capacitor, which is not resisting the current flow at all. A quarter of a cycle later, when the current gets going up to full speed in the inductor and it no longer offers any inertial resistance to that current, the voltage in the capacitor has built up to a maximum, so it blasts that high current through the inductor. The overall result is that the pair of devices acts like it has *no resistance* at all, but just at that one *resonant frequency*. It becomes a short circuit. (Note that this is only true if there is pure capacitance and pure inductance, and there is no ordinary resistance. Of course, in any real circuit, there is always some resistance in the wires, etc., and this decreases the effect somewhat.)

Mathematically, the formula in the optional section on page 125 shows that the two reactances cancel each other if they have equal magnitudes, leaving only the ordinary resistance, R. This is a clue that the *resonant frequency*, f, occurs at whatever value $X_C = X_L$. Plugging in the formulas on pages 99 and 113,

$$X_C = \frac{1}{2\pi f C} = X_L = 2\pi f L. \qquad (11.7)$$

Rearranging,

$$f = 1 / (2\pi \sqrt{LC}) \qquad (11.8)$$

In the *parallel* case of Fig. 11.5, the impedance (effective overall resistance to ac voltage) is a *maximum* at that same resonant frequency, and if a wave of that frequency were coming along in the wires, it would not be short circuited. If there was no ordinary resistance in the wires, that wave would start an oscillation, going back and forth *within* the square that makes up the parallel circuit, and it could continue for a long time. (Note, however, that it has to be *started* by something else.) It would reach its peak voltage at the same instant that the wave coming in from the left reaches its peak. Therefore, as electronic engineers say, "it would look like an *infinite resistance*" to the incoming wave. (The square that makes up a parallel "resonator" like this is cometimes called a "tank circuit.")

COMMENT

Resonant circuits are often used as filters. The *parallel* type can be used to short out everything except one particular frequency, and it is often used this way in radio receivers and transmitters. In the *series* type, if the inductor is the *primary* coil of a transformer, then only one frequency will come out of the *secondary* coil (not shown in the figure).

The *more* energy that is lost through heat, via the electrical *resistance*, the sooner the oscillation will die out — also, the *less sharply* the oscillation will be "tuned" to one narrow frequency. By comparison, *lower resistance* provides *sharper* tuning — higher Q (or "quality factor," as mentioned on page 101).

EXPERIMENTS

Parallel LC Resonator

The experimenter can assemble the circuit shown at the upper left of Fig. 11.5. The 1,000 mfd capacitor is used, and the "12 volt" secondary of the transformer is the inductor, using just one yellow wire and the center-tapped black wire.

Then attach the vertical input of the oscilloscope to the resonator output. The scope can be set as on page 122, except that the TRIGGER SOURCE should be VERT, and the TIME/DIV should be 0.1 s (a tenth of a second per square division on the screen). The channel 2 (Y) sensitivity can be about 2 V/DIV. Place the bright spot (or line) in the middle of the screen, using POSITION [Y].

A nine volt battery (not shown in the diagram) is momentarily attached to the resonator input. On the scope, a bouncing sort of action is observed, in which the bright spot flies upward and then down, but it goes down *past* the middle of the screen and then is damped back up to the middle again. (The current flow in the inductor has continued to push the voltage down below zero, becoming negative.)

The capacitor has a small amount of inductance, and the inductor a small amount of capacitance, so either one alone will also show some resonance, but not as much as both together. If the 120 volt primary coil is used instead, there is not as much action, because not enough current flows to build up a strong magnetic field. (Note: Resistor-capacitor resonators are similar to LC types.)

EQUIPMENT NOTES

Components   Radio Shack Number

Component	Radio Shack Number
Battery, nine volts	23-653 or 23-553
Clip leads, 14 inch, one set.	278-1156C or 278-1157
Resistor, 100K, 1/2 or 1/4 watt	271-306 or 271-308
Capacitor, 1000 mfd	272-1019 or 272-1032
Transformer, 120V to 12V, 450 ma, center tapped	273-1365A or 273-1352

Oscilloscope, Model OS-5020, LG Precision Co., Ltd., of Seoul, Korea

CHAPTER 12

Relays

WHAT THEY ARE

The ordinary type of relay is an electromagnet that operates a switch. Many kinds of switch can be used, including the "SPDT" (see Index to find explanation in an earlier chapter), which is shown in Fig. 12.1. In this course, the inexpensive relay that is recommended is a "DPDT," but its full capabilities will not be exploited, and it will just be operated in the single pole mode.

Relays are analogous to water faucets, because both allow small amounts of energy to control very large amounts of energy. Relays are in use all around us, for everyday applications such as controlling air conditioners and heating systems, and for limiting electric currents by means of circuit breakers. For many decades, special forms of relays such as "stepping switches" and "crossbar systems" were the main devices used to automatically switch telephone signals, thus selecting which phone was to be connected to another. A disadvantage of the relay, however, is that it can only be all the way "on" or all the way "off," and not in between.

In the circuit of Fig. 12.1, the 15 ma at 10 volts in the coil could be used to control as much as 15 amperes at 100 volts in the contacts, a power increase of 10,000 times. The relay itself includes the parts within the dashed-line rectangle. Small open circles on that line are the "terminals" of the relay. "Electronic technicians and engineers" (people working with small amounts of electricity in fairly complex circuits) ordinarily use the symbols shown in the top half of the figure.   However, "electricians" (people working with large amounts of electricity, usually with less complex circuitry) have their own set of symbols, shown in the lower diagram for the same circuit. In electrician's notation, the relay coil and the contacts are not necessarily drawn near each other, but they are labeled in some identifiable way.

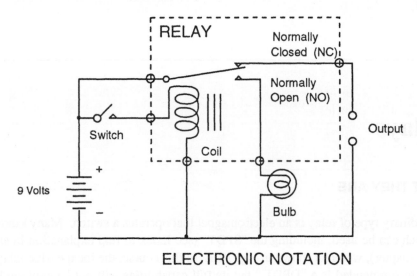

ELECTRONIC NOTATION

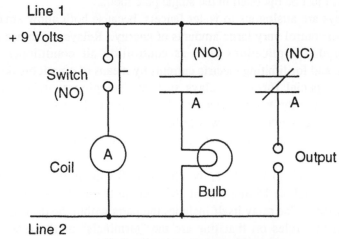

ELECTRICIAN'S NOTATION
(Sometimes called a "ladder network")

Fig. 12.1 A basic relay circuit, shown with both types of symbols.

Electricians sometimes use the other symbols shown in Fig. 12.2. Also, these are fairly common in the electronics diagrams used in countries other than the U.S.A. The reader is encouraged to use the Index of this book, under the word "symbols," or under more specific words such as "ground," to find other frequently used ways to draw components.

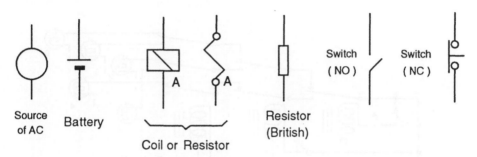

Source of AC Battery   Coil or Resistor   Resistor (British)   Switch (NO)   Switch (NC)

Fig. 12.2 Some alternative symbols used by electricians and others.

Because relays can be *only on or off,* telephone engineers at AT&T Corp. brought the relay-compatible "binary" numbering system (using *only zeroes and ones*) to a high level of development, in order to control nationwide relay-operated switching systems. Those telephone switching circuits were really specialized computers, even back in the days when the word "computer" had not yet appeared in English language dictionaries. Some of the old relay switches are still in use in small towns, connected to modern transistorized switches in big cities, but old or new, they all involve binary numbering.

In binary numbers, instead of counting upwards from 1 to 2, the counting has to jump all the way to 10, since there are no twos available, only zeroes and ones. Then, the third number is 11, and the fourth has to go all the way to 100, and so on. This is the heart of "digital" electronics. (More aspects of digital systems will be explained in later chapters. Simplified examples of digital computation and machine language will be included in the last experiments of this course.)

Even though modern *transistors* can also be set to any voltage *in-between* zero and one, there are still great advantages to the binary system, such as its vastly decreased susceptibility to random errors ("noise"). Therefore binary is used just as much in modern equipment as with the old relays. However, the decimal numbers that humans understand must be "translated" into binary, so the machines can use them, and then back into decimal again later, so that human users of the system can understand the results. (See also Chapter 19, Fig. 19.6.)

The circuit of Fig. 12.3 illustrates how a relay circuit can translate binary numbers back into decimal, at least up to the number 4. The table below it shows the translation. For example, if relays B and C are activated by switches, then light bulb 3 will glow. This is not a practical circuit, but it illustrates the general idea of how some types of translators work.

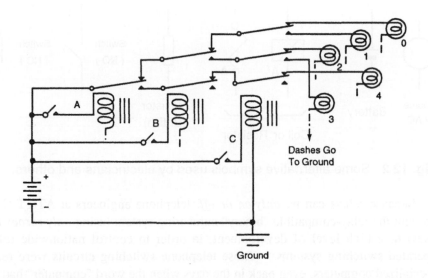

Fig. 12.3 A relay-operated translating machine, for binary to decimal.

In the following table, the reader should compare the left-hand column with the middle column and note that "A" controls the first binary digit, and so on.

Activated Switches and Relays	Corresponding *Binary* Number	Corresponding *Decimal* Number and Lit Bulb
None	0 0 0	0
C	0 0 1	1
B	0 1 0	2
B & C	0 1 1	3
A	1 0 0	4

EXPERIMENTS

Simple Relay Circuit

The circuit of Fig. 12.1 (page 130) should be constructed. Even though both the relay and the bulb are designed to work at 12 volts, they will operate satisfactorily at 9.5 volts. However, if the battery has been used a lot and only provides about 8.5 volts, two batteries might have to be put in series to do these experiments. (The doubled voltage will not damage these components.)

It might also be noted that the relay will not operate on the 12 volts of ac available from the transformer, because the relay's inductance does not allow sufficient ac current to flow. (If a transformer is available, the experimenter might try this as an optional item.)

COMMENT

An important point for the reader to note here is that a relay designed for use with 12 volts of *ac* would overheat if driven by 12 volts of *dc*, because too much current would flow if not held back by *inductance*. Similarly, other ac devices having coils should never be driven at the nominal voltage with dc.

Once again, the "switch" in Fig. 12.1 is really just an unattached clip lead. When it does become attached to the battery, the relay makes an audible clicking noise and the bulb lights. If the relay is the type with a transparent plastic case, the experimenter should observe the soft iron "armature" rotating slightly around its stationary hinge, with the NC contacts opening, followed very quickly by the NO contacts closing.

Timing with the Oscilloscope

The various times involved in a relay's operation can be measured with an oscilloscope, even though they are just thousandths of a second long. The circuit is shown in Fig. 12.4. An internal circuit diagram of the relay itself is printed on its case. The resistor and capacitor hooked up by the experimenter will slow some parts of the operation for easier observation on the scope.

The scope should be set *similar* to the way it was done on pages 76 and 77. For example, the MODE for input will be CH2, and the MODE for triggering will be AUTO. A bright line appears across the screen with the AUTO setting, as long as there is not much voltage coming into channel 2. As soon as significant voltage appears, the AUTO mode internally stops triggering the horizontal sawtooth (page 73), and the bright spot stops moving, while the scope waits for some other "trigger" action to start it moving horizontally.

The POSITION Y knob can be used to move the bright line up or down to the middle of the screen. The POSITION X knob can adjust the starting point of the line, so it is barely to the right of the left edge of the screen.

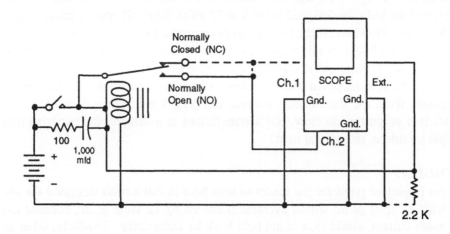

Fig. 12.4   Measuring the speed of relay operation. (In some cases the 2.2K might be needed, in order to discharge a capacitor that is in the scope's "external trigger" circuit, before each new timing operation.)

Following are some things that are different from those on pages 76 and 77.

A second BNC cable is attached to the EXT TRIG socket, for external triggering. (The dashed- and dotted-line wires in Fig. 12.4 are not attached yet.)
TRIGGER SOURCE should be set to EXT.
TRIGGER LEVEL knob should be slightly positive, about 1:00 o'clock.
TIME / DIV should be 5 ms.

Then the TRIGGER MODE switch is moved from AUTO to NORM (normal), and the bright line disappears. The scope is waiting for an external trigger signal to come in through the EXT. BNC cable.

The experimenter should now momentarily close the "switch" (really a clip lead connection) in Fig. 12.4. This applies 9 volts to the relay coil, and also to the EXT. terminals of the scope. The latter causes the bright spot to start moving across the screen, towards the right, at a speed of one large division (a centimeter) per 0.005 second. About 15 ms later (approximately 3 squares to the right), the 9 volts appears in the CH2 cable, raising the bright spot about 2 squares. This means that the relay's armature has moved down and hit the normally open contact at that time. The horizontal bright line then continues, giving the appearance of the "step function," which is line A in Fig. 12.5, below. (Upon removal of the battery's 9 volts, another pattern might appear, which can be ignored in these experiments.)

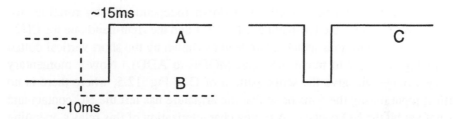

Fig. 12.5 Oscilloscope patterns that appear during these experiments.

The experiment can be repeated quickly, while adjusting the TRIGGER LEVEL knob to higher values (above the available 9 volts), until it stops the triggering action. It can then be set to a slightly lower value ("backed off") for the next part of the experiment.

The TRIGGER SOURCE is moved to VERT, and the BNC cable is then disconnected from the EXT socket and from the relay. Now the incoming signal that triggers (starts) the bright spot moving is whatever comes into *either* the CH1 or CH2 sockets for BNC cables. (If the setting had been CH1 instead of VERT, then *only* a CH1 signal would trigger the spot motion.) When the battery voltage is repeatedly applied, on and off, only a straight horizontal line appears (not shown in Fig. 12.5). The reason is that the triggering does not occur via CH2 until the relay has already operated, and then nothing else happens (except maybe some contact bounce).

The next experiment is to plug in the BNC cable to the CH1 socket, and attach its hook to the relay's NC contact, as shown by the horizontal dashed-line in the figure (not the short vertical dotted line). Now when the 9 volts is momentarily applied again, the step function A reappears, because the first voltage comes in through the NC contact, which lasts long enough to trigger the sweep motion. The vertical input (CH2) does not occur until the NO contacts have closed, just like in the EXT trigger experiment above.

Moving the TRIGGER SOURCE slide switch to CH1 does the same thing, because CH1 is still the first triggering input. (The *vertical input* is still CH2.)

Changing the vertical input MODE from CH2 to CH1 makes the screen pattern change to B in Fig. 12.5, because it takes about 10 milliseconds for the NC contact to open, being somewhat delayed by the resistor and capacitor. (The capacitor takes about that long to charge up to 8 volts or so, which then can operate the relay.) The TRIGGER SOURCE can now be moved back to VERT.

Sliding the input MODE switch to DUAL, and repeating the momentary voltage, the scope pattern shows *both* the solid line A and the dashed line B of Fig. 12.5. Either can be moved up or down independently by rotating two POSITION Y knobs, the left-hand one for CH1 and the right-hand one for CH2.

The last thing to do is attach a clip lead as shown by the short vertical dotted line in Fig. 12.4 (or to move the input MODE to ADD). Now a momentary battery voltage will give the scope pattern of C in Fig. 12.5, where there is no vertical input during the 5 ms or so that the armature has left the NC contact and has not yet hit the NO contact. A timing characterization of this relay's operating modes (with the resistor and capacitor attached) has now been completed.

Positive Feedback and Latching

The circuit of Fig. 12.6 illustrates "positive feedback," and it can also be used to construct a "circuit breaker." The experimenter can use the 5K pot as a rheostat in this hookup, beginning by using the full 5,000 ohms capability (with the arrow

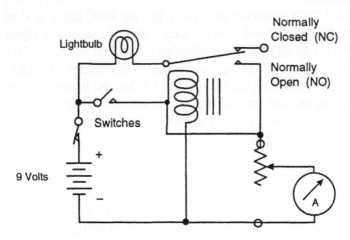

Fig. 12.6   A latching relay circuit, used as a circuit breaker.

all the way downward in the diagram). The meter is used in its milliamperes setting. The "switches," as usual, are just clip leads. The normally closed "switch" (vertical in the diagram) is a clip lead attached to the battery.

When the normally open clip lead "switch" (horizontal in the diagram) is momentarily closed, the relay "pulls in" (operates), and the relay contacts then start sending current to the coil, so that the coil stays energized even after the horizontal "switch" opens again.

The phrase "positive feedback" means that, when a device like a relay operates, a signal of some sort is sent back to the beginning of its circuit, to continue its operation, or make its operation stronger. Another description of this circuit is "latching relay," although that can also mean a strictly mechanical means of keeping the armature down, via a ratchet.

Circuit Breaker

If the "output" of this relay (passing through the two small circles) was an electric heater or motor, represented in this diagram by the rheostat and ammeter, then it could be used as a "circuit breaker," to limit the current and prevent damage in case of a short circuit or other excessive load current. In this experiment we will imagine that an "excessive" current is 35 milliamps. The experimenter can slowly decrease the resistance of the rheostat, observing the ma readings. When the current gets as high as about 25 ma, the relay will click and cut off the output.

The lamp bulb has a resistance of about 100 ohms when it is partially lit, and the relay coil also is about 100 ohms with dc current flowing. If the battery is fresh and delivering a little less than 10 volts, then the "pull-in current" is a little less than 50 ma, which is sufficient to initially operate the relay.

Once the armature has moved to the pulled-in position, it is closer to the electromagnet, and it "completes the magnetic circuit" better than when it was previously in the initial "released" position, and it has more attraction to the iron armature. Therefore, less current is required to continue the pulled-in position. Now the coil current would have to be decreased down to about 30 ma in order for the relay to click off again, and this is called the "release current" or "drop out current." When the "output" current to the meter gets to be about 25 ma, that plus the 50 ma through the relay is roughly 75 ma. Going through the 100 ohms of the bulb, that will cause a voltage drop (page 41) of about 7.5 volts, leaving less than 2.5 volts across the relay coil. With a coil resistance of about 100 ohms, less than 25 ma will flow, and the relay will release.

Commercially available circuit breakers ordinarily have more than one coil, with the extra one being a separate relay that can cause the main one to operate, disconnecting the power. Other designs are also used.

Negative Feedback and Oscillation

In the circuit of Fig. 12.7, there is an action that "feeds back" and makes the operation of the relay weaker, rather than stronger. This action is simple and direct — it is the disconnection of power to the coil. In the first experiment, the capacitor is not attached. A high pitched buzz is heard, because the armature vibrates (oscillates) at a fairly high frequency. The negative feedback occurs *too soon* for the armature to get all the way down to the NO contacts.

This type of simple circuit (without the neon bulb or capacitor) is commonly used to operate doorbells, with a bell clapper attached to the armature, which hits a metal bell. It was also used for the older model telephone "ringers," before modern transistorized oscillators became available. Without the clapper and bell, it is still used a lot for various "buzzer" applications. (More modern buzzer designs use the piezoelectric diodes discussed in Chapters 14 and 15.)

If the relay has a transparent case, sparks can be seen at the NC contacts, and this would eventually damage them, if the experiment were continued for a long time. The inductive kick of the coil is not allowing the current to stop very fast.

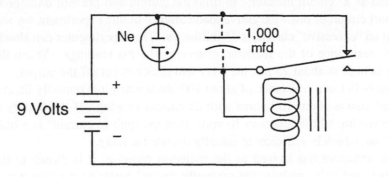

Fig. 12.7 An oscillator with negative feedback.

Attaching the capacitor with clip leads as shown by the dashed lines, the sound changes to definite clicks, separated by a second or so. This is complete oscillation, which is more effective because the capacitor has delayed the current stoppage until there is enough time for completing the operation. If the capacitor is used, and if output wires (not shown) were attached between the NO contact and the negative wire, then the output signal would be a "square wave."

The lesson here is that negative feedback is more effective in causing oscillation if it is *delayed somewhat,* depending on the type of operation desired. (This will be seen again in Chapter 16, using *transistors.* However, *positive* feedback can also lead to oscillation with transistors, and in fact it is the usual method, with very fast devices like transistors or vacuum tubes.)

Another effect of the capacitor is that it acts as a snubber (page 110), absorbing the inductive kick and preventing the neon bulb from flashing. Of great importance is that the snubber prevents the spark and makes the contacts last far longer. A small resistor, about 10 ohms, could also be put in series with the capacitor to protect it, but it is not needed for just a short time of use. A much larger resistor would interfere with the desired action here, since the available voltage is barely high enough to operate this relay. However, as an optional experiment, the incandescent bulb could be wired in series with the capacitor, to see if the oscillator still works — it probably will work, if a fresh battery is used, supplying 9.4 volts, or if two batteries are wired in series, but probably not if the battery has been drained down to only about 8 volts output.

A better snubber, and one which is more often used in industrial circuits, is a semiconductor diode, to be explained in Chapter 14. In fact, an experiment making use of a diode snubber is described on pages 159 and 160.

COMMENT

The transistors which will be discussed in Chapter 15 can operate about a million times faster than relays (nanoseconds instead of milliseconds). Also, they can be made about a million times cheaper if they are part of an "integrated circuit," and they can be made about a billion times smaller in volume, compared to relays. A relay is very reliable for a few million operating cycles, but a transistor's cycle life is unlimited. For these reasons, relays are being used less, and transistors more, in modern equipment.

In spite of these differences, relays are still valuable components, because they can handle large voltages and currents, and it is easy to design circuits with them. Also, they are very common in older equipment, much of which is still in use and unlikely to be replaced soon.

Because of the spectacular improvements in manufacturing methods for producing semiconductor integrated circuits, it is cheaper to make an "electronic relay" composed of many transistors (sometimes a hundred or so) than to make a single "electromechanical" relay with its coil and moving armature. Therefore, newer electronic equipment tends to be designed with minimal moving parts such as armatures, even if the design has to be far more complex. About ten years ago, these electronic relays were rather easily damaged, but new designs are protected by special semiconductor diode snubbers (including "varistor diodes" and "Zener diodes"), and now they are more reliable than the electromechanical types.

EQUIPMENT NOTES

Components For This Chapter	Radio Shack Catalog Number
Battery, nine volts	23-653
Clip leads, 14 inch, one set.	278-1156C
Plug-in Relay, 12VDC, DPDT	275-218 or 275-248
Lamp bulb, tungsten, 12 volt dc, green or blue	272-337A or B or C
Resistor, 100 ohm, 10 or 1 watt	271-135 or 271-152A
Resistor, 2.2K (1/2 or 1/4 W)	271-306 or 271-308
Capacitor, 1000 mfd	272-1019 or 272-1032
Oscilloscope, Model OS-5020, LG Precision Co., Ltd., of Seoul, Korea	
Potentiometer, 5K	271-1714 or 271-1720
Multimeter (or "multitester")	22-218 or 22-221
Neon test light, 90 volts	22-102 or 272-707

CHAPTER 13

Semiconductors

WHAT THEY ARE

While most *metals* like copper and silver are electrical conductors, most *oxides* like quartz and sapphire are insulators. "Semiconductors" have conductivities that are in between, examples being a few metals like selenium and a few compounds like copper oxide. In conductive metals, the electric current consists of moving electrons, but in semiconductors it can be *either* negatively charged electrons, or something else: positively charged "holes."

If a semiconductor having mostly electrons is put in contact with a different kind having mostly holes, the electricity can be allowed to go in one direction but prevented from going in the opposite direction. Thus ac can be converted to dc, and this is called "rectification" (left over from the days when ac was considered disadvantageous, and converting it to dc was making it "right" again). The device that is made this way has two wires attached to it and is called a "diode" ("di-" meaning two in Latin, and "-ode" standing for "electrode").

If *three* semiconductors are put together in just the right way, a small current through one can control a large current through the others ("amplification" of the current). The resulting device is called a "transistor," (short for "transfer resistor").

Quantum Mechanics

The reader probably remembers from chemistry courses that the electrons orbiting around in the outer shells of an atom can only have certain energies, but not others. Nature allows these "energy levels" to be whole numbers of the lowest energy units, such as 2.0 times the lowest level, or 3.0 times, but *not* 2.1 or 2.2.

A complex set of mathematical relationships called "quantum mechanics" seems to describe the behavior of electrons quite truthfully. It tells us that some numbers are favorable, and others are not. In some ways it is like the ancient superstition of numerology, where 7 might be lucky and 13 unlucky. The difference is, however, that in the long run, betting on 7 in roulette and avoiding 13 will *not* benefit the gambler, but following the rules of quantum mechanics *does* predict new chemical medicines and new electronic devices very successfully and profitably. The ability to predict things that have not yet been made, and then to be able to make them, is an important criterion of "truth" in science.

It matters little whether we can fully understand a new scientific principle, or whether we like it and feel comfortable with it. What matters most is whether we can use it to make valuable things (without doing any harm, of course). Programs that will operate in a desktop computer can now use quantum mechanics calculations to design new medicines and new microwave semiconductors for cellular phones, things which are obviously of great value.

As far as understanding quantum mechanics is concerned, the math has now advanced to the uncomfortable point where some of its principles say certain conclusions *have to be* true, and others say they *can't be* true, simultaneously.* In other words, human beings don't seem to be able to "understand" this relatively new science. However, we certainly are making good use of it, without a full understanding.

Energy Bands

The most likely energy of an outer electron in a silicon atom is something like a tuned guitar string, or a pendulum. The electron obeys many of the same natural laws that a much bigger object does when vibrating — that is, some energies are "allowed," like the second harmonic of a plucked string, and others are "forbidden," like a musical tone that is 2.1 or 2.2 times the natural tone of a string.

---

* (a) John Horgan, 'Quantum Philosophy: The Surreal Realm,' *Scientific American* Vol. 267 (July 1992) pages 94-99.
 (b) John Horgan, *The End of Science* (Addison-Wesley Publishing Co,, 1996).

Similarly, a pendulum with a soft, flexible shaft has one natural frequency of oscillation, and several definite multiples of it, if the shaft is bending. However, as shown in Fig. 13.1, if two pendula are loosely coupled by a soft spring (or two guitar strings by a piece of rubber), there will still be a most-likely natural frequency, but two "side-band frequencies" will appear.

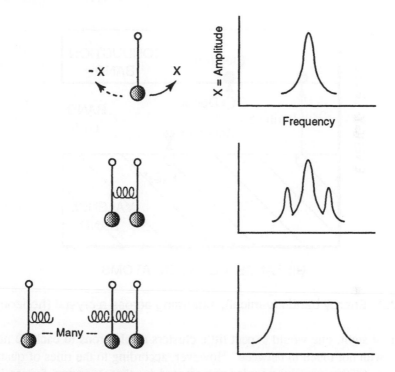

Fig. 13.1 Coupled pendula, vibrating at their natural frequencies.

An important thing to note is that if a million pendula are coupled, there will be a fairly wide band of frequencies, and so many other frequencies would be inside this band, that they would blend together. This is shown in the figure as a horizontal band. Higher frequencies have higher energies, so the amount of *energy* increases *horizontally* toward the right.

Band Gaps

The way we usually graph the *energy* of an electron is on the *vertical* axis, as in Fig. 13.2. The horizontal axis is the distance from the left-hand edge of a crystal of pure elemental silicon. We are not considering the nuclei of the silicon atoms here, with their protons and neutrons, but only the outer electrons of the atoms.

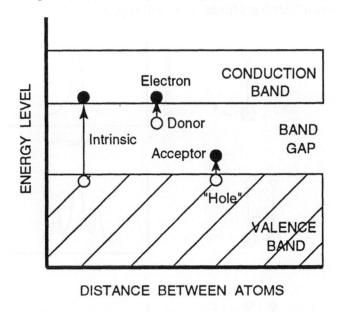

Fig. 13.2 Energy bands (vertical), extending across a crystal (horizontal).

At first sight, one would expect little clusters of electrons to each be near a nucleus, with not much in between. However, according to the rules of quantum mechanics (discovered by people, not created by them), nature favors large groups of atoms being tightly linked together, in a repeating (called "periodic") structure. When this occurs, as in a nearly perfect single crystal, the atoms are coupled, like pendula attached to springs. Instead of springs, there are chemical bonds, which form because of the rules of quantum mechanics. Not only do the energies merge into bands (shown vertically here), but they merge horizontally also. Thus an electron could move freely from atom to atom (horizontally here), existing at places *in between* the atoms, just as easily as *near* an atom. This is part of the reason why electrons can travel freely through a metal wire, even though the atoms are discontinuous, with empty spaces between them — the electrons move continuously, like a liquid or a gas in a pipe.

There is another band, at higher energy, called the "conduction band." It has no electrons in it, at least with the element silicon. In between is the "energy gap," which is sometimes called "the forbidden band," where nature's rules do not allow any electrons to have their energies.

The electrons in the bonds that hold the silicon atoms together are shown in the "valence band." One would expect those electrons to be able to move, all together, if there was a voltage applied to the crystal (from a battery, for example). However, a highly perfect single crystal of very pure silicon is a *fairly good insulator,* rather than being a conductor of electricity. The reason, explained by quantum mechanics, is that the electrons fill all of the allowed horizontal spaces, so the only way they could move is if they all move together, at *exactly* the same time. Quantum mechanics tells us that nature insists on some randomness, and the probability of all the electrons moving together, *exactly,* is so low, that it essentially never happens.

Metals

A logical question is, "Then how does silver metal conduct electricity?" The answer is that the conduction band in a metal is much lower than shown in Fig. 13.2, and in fact it overlaps the valence band, without any gap. At various places in the crystal, thermal energy that is available at room temperature can supply very small upward kicks to individual electrons in the valence band, so the electrons go into the conduction band. Then those particular electrons are free to move horizontally from atom to atom, whenever an electric field is present, as from a battery attached to each end of the metal crystal. There is plenty of room in the band for the electrons to move horizontally at *different* times, without bumping into each other. Therefore the metal conducts electricity.

Insulators

Crystals of quartz (silicon dioxide) and sapphire (aluminum oxide), to name two *very good insulators,* have large band gaps (both around 8 electron volts, or "ev"), so they do not conduct electricity at room temperature, for the same reason that pure silicon does not. However, if they are *heated* to about 750°C (red hot), a few electrons from the valence band will randomly have the energy required to put them up in the conduction band, and the materials will become slightly conductive. In the figure, this is represented by the "intrinsic" arrow. Silicon, with an energy gap of about 1 ev, only needs to be heated to about 200°C in order to conduct via the intrinsic mode.

Very high electric fields, such as 1,000 volts across a crystal only 10 microns thick, or 100 volts across 1 micron, can do the same thing as raising the temperature: cause intrinsic conduction in silicon *dioxide* ("sili*ca*," see page 267). In pure ("intrinsic") sili*con*, it only takes a *fairly* high electric field to cause intrinsic conduction at room temperature.

When an extremely high electric *field* does cause conduction, the flow of current can *heat* the material, thus thermally promoting more electrons into the conduction band, and making still more conduction. This is a "vicious circle," where the more it happens, the more it happens. It is called an "avalanche," and if it continues, the material might be destroyed by the heat, which is called "dielectric breakdown." (See also "avalanche" on page 19.) Some materials such as mica tend to stand up to high fields without breakdown, and they are said to have high "dielectric strength." That is not at all the same, however, as the "dielectric constant" described on page 92, which refers to an increase in capacitance, although the two phrases are sometimes confused with each other.

Light is also a form of energy ("photons"), of course, and it can also cause an intrinsic semiconductor to become a conductor. This is called "photo-conduction," and it is used in photocopiers and laser printers (page 247).

Dopants
Semiconductors such as silicon (Si), with only medium sized energy gaps, can easily be made into *fairly good conductors.* As the reader might remember from chemistry courses, silicon has a valence of four, similar to carbon, germanium, and tin. If an element such as phosphorus (P) or arsenic (As), with a valence of 5, is somehow put into a silicon crystal, quantum mechanics predicts that it will take on the valence of the rest of the crystal, so it only makes 4 bonds to the surrounding Si atoms. This leaves one *extra* electron in its outer shell, which is not used in any bonds, and is therefore easily taken away.

Those P or As atoms have energy levels inside the Si band gap, shown in Fig. 13.2 as the spot labeled "donor." The small amount of random energy available at room temperature is enough to kick some of their *extra* electrons up into the conduction band, where a horizontal electric field (at least, "horizontal" as shown in this diagram) can cause it to move anywhere throughout the crystal. As long as the conduction band is not 100% filled up, the material then becomes a conductor.

It only requires about 0.0001% ("one part per million") of P or As in Si to convert a fairly good insulator into a fairly good conductor. The P or As are the "donors" in the "semiconductors." The act of putting a small amount of donor into an otherwise intrinsic semiconductor is called "doping." It is often done by heating the pure Si in contact with some donor material to about 1,000°C, in which case the donor diffuses into the main crystal.

A semiconductor that has been doped so that electrons (negative "charge carriers") can conduct is called "N-type" material. The original pure material was "I-type," before being doped.

Boron (B) or aluminum (Al), having main valences of only 3, can also be diffused into Si, causing them to "accept" an electron from the Si crystal, instead of donating one. This is also shown in the figure. What is left behind is a *lack of* one electron in the Si, and this is called a "hole." An electron in the valence band can now move into that hole, in response to an electric field (from a battery, etc.), leaving a new hole next to the old one. Then another electron can move into the new hole, again making an even newer hole nearby. Thus the valence electrons do not all have to move at the same time, so they are free to conduct, where before doping they were not. (Note, however, that the electron that was accepted by the B or Al is *not free to move,* since there is no repeating B or Al crystal structure nearby.)   For every electron, there is a positively charged proton in the nucleus of the atom, which is not shown in the figure. Thus if an electron moves to the right, there is a positive charge which *acts as though* it is moving to the left. The protons do not really move, but for simplicity's sake, *it is imagined that a positively charged hole is moving,* when really it is just the electrons that move (in the opposite direction).   Silicon doped this way is called "P-type" material.

Looking back for a moment at the N-type material, the donor leaves a new hole at its location, which *can not move.* A moving electron can fall back into that hole, and is then temporarily "trapped," until it gets enough random heat energy to jump out again.   Similarly, a moving hole can fall back up into an acceptor location and be trapped, and it does not take much energy change for it to get out again.   These facts are important in understanding the various interactions with light, as will be explained further on page 244.

Heating a semiconductor causes more conduction, because more electrons can go upward and be donated or accepted. However, there is an opposite effect, which is less powerful. Heat causes the *atoms* to move around randomly, so they are no longer in their exact "periodic" positions (see last paragraph on page 144). Thus the quantum mechanics rules are not being followed so well, and the charge carriers (electrons or holes) can not move as easily (their "mobility" is less). This is called "scattering" of the carriers.

In metals there are enough carriers without heating, so scattering causes a *decrease* in conductivity when the temperature is raised.   That is why the resistance changes when a light bulb is turned on (pages 19 and 84).   Both phenomena, the increased resistance of metals when heated, and the decreased resistance of semiconductors, are used in "sensors" that can measure temperature.

CHAPTER 14

Diodes

WHAT THEY ARE

As mentioned on page 141, a diode is an electronic device having two terminals. There are many types, but the most common one is the PN diode. This can "rectify" ac, by only allowing current to flow in one direction, thus converting the ac into dc.

The P-type semiconductor (usually single crystal silicon, doped with B or Al) is contacting N-type material (usually Si doped with P or As). This can be made from a piece of intrinsic (very pure) silicon, with acceptors diffused into half of it and donors into the other half, ending up with P-type making contact to N-type along a plane inside the crystal. The contact plane is called a "junction."

Alternatively the PN junction can be made by growing the *whole* crystal with a moderate amount of *acceptor* material already mixed into it, and then diffusing a relatively larger amount of *donor* atoms into *half* of it, overwhelming the effect of the acceptors in that half. Again, where the two types meet there is a planar junction.

Rectifiers

If an external voltage, from a battery for example, is put across a PN diode so the P type material is made positive and the N-type negative, then current will flow, as shown in Fig. 14.1. The electrons "fall into" the holes, becoming lower energy electrons. The energy they lose can be emitted as light ("photons") and heat (random vibrations of the silicon's atoms). A silicon diode does not produce visible light, but if gallium phosphide is used instead of silicon, this effect can be optimized so the emitted light is visible.

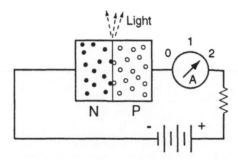

Fig. 14.1 A PN diode, "forward biased," with current flowing. Black dots are electrons, and circles are holes. (A "bias" is a small voltage applied to a device, tending to make a current flow in a certain direction.)

It is interesting to note here that if there is no battery, but just an ammeter in the circuit, and light shines on the diode junction from some external source, then a small electric current will flow, but in the "reverse" direction, with electrons coming out of the N-type material and going through the ammeter into the P-type half of the diode. This is the principle of the "solar cell," and an experiment on the phenomenon will be described in Chapter 23.

Now let us suppose that the battery polarity is reversed, and + is attached to the N-type half of the diode, as in Fig. 14.2. The current will be close to zero, and for practical purposes we can consider the diode to be an insulator. The reason is that both the electrons and the holes are attracted *away from* the junction, leaving what is called a "depletion zone," or "depletion region." This material has no "charge carriers" in either the conduction band or the valence band, so it is an insulator. All the electrons in the valence band, within that volume of silicon, would have to move at exactly the same time, so they essentially don't move at all.

If the temperature is high enough, or the electric field (proportional to voltage) is high enough, some intrinsic conductivity can occur (bottom of page 145), and in fact this could lead to "avalanche" (page 146). In most cases we do not use such high fields, because they might lead to what is considered "failure" of the diode. However, some devices are purposely designed to make use of avalanche, and they are called "Zener diodes," named after the person who studied the scientific basis of this design. They will be described in more detail later and actually used in some of our experiments.

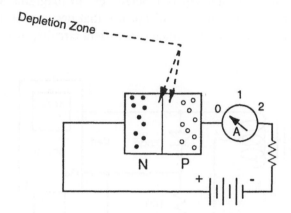

Fig. 14.2 A "reverse biased" PN diode, with nearly zero current flowing.

EXPERIMENTS

Characteristic Curves

The first thing to do with a rectifying diode is test it, by attaching the ohmmeter to its two terminals. The white or silver ring printed on the black exterior "package" of the diode is near the N-type silicon, so the *ohmmeter* probe that has electrons coming *out* of it (red, with most inexpensive ohmmeters) can be touched to that terminal, or a clip lead can be used to make contact to it. (If the package is transparent glass, then the ring is black or very dark red.) When the other ohmmeter probe touches the other diode terminal, the meter will indicate a low resistance, provided that the diode is not defective. (Inexpensive diodes sometimes are defective, so they should be tested before use.) This test is *not* shown in the figures, because it is simple enough to describe without a diagram.

In Fig. 14.3 on the next page, the diode is shown as an arrow-like triangle, with a short line across its point. That short line corresponds to the white or silver line printed on the actual diode. In the symbol of the figure, the short line shows the terminal where negative electrons could flow into the device, while the arrow's triangle shows the *direction* that "positive current" (which is really just imaginary) could flow through the device.

The next experiment is to determine the "characteristic curve" of a rectifying diode. The curve tracer of Fig. 14.3 is assembled, with the diode attached as shown. The 6 volt center tap of the transformer is used to provide *ac* voltage, and this way we can see what happens to the diode when the voltage is applied in each direction. (Sometimes a center tap is labeled "ct" in diagrams and also in component supplier catalogs, but we will not use that designation here.) The oscilloscope is set up as in previous voltage-vs.-current circuits, such as the ones on pages 84 and 97.

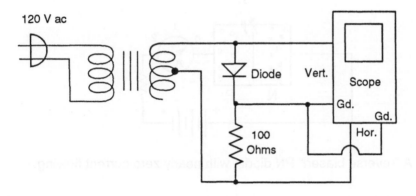

Fig. 14.3 Curve tracer, to study diode (or other "DUT") behavior.

The type of pattern that is seen on the scope is illustrated at the upper left of Fig. 14.4. When the "top" of the transformer secondary coil is positive (and of course, the center tap is then negative), current flows, which puts a voltage drop across the 100 ohm resistor. The horizontal display thus shows that current is flowing, but there is very little voltage drop across the "forward biased" diode, so the vertical display shows very little displacement above its center. However, when the top of the coil becomes negative, it is just the opposite: a lot of negative (downward) voltage is shown, and practically no current. (Whatever is being tested — the diode in this case — is called the "device under test," or "DUT.")

The experimenter should adjust the POSITION Y and POSITION X knobs on the scope so the image is near the center of the screen. Also, the VOLTS/DIV and both VARIABLE knobs can be adjusted to bring the whole picture into view.

Note for comparison, that a resistor is simply linear in both directions. (The experimenter might confirm this, as an optional experiment, by putting a 150 ohm resistor in the circuit to replace the diode.) Also, the curves of the inductors and capacitor and light bulb are reviewed in the figure (see index for previous pages).

It should be noted that there is an appreciable forward voltage drop, about 0.6 volts when the current is around 1 ma. This is typical for silicon PN diodes, but diodes made from the element germanium can be made with "forward voltages" of about 0.3 volts. Completely different kinds of diode, not involving PN junctions can be made, such as Schottky diodes, and some have even lower forward voltages. (The nonlinearity at very low current is "crossover distortion.")

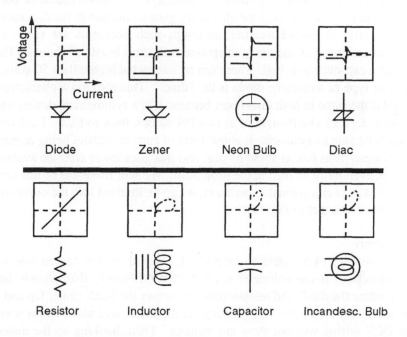

Fig. 14.4 Curve traces for various kinds of diodes.

Speaking of different kinds of diode, there are many, and a few are shown here (and more on page 161). The experimenter should put a 6.2 volt Zener diode in the curve tracer and confirm the display in the figure. This is a device that is specially designed to "go into avalanche" (bottom of page 150) at a known voltage, without damaging the device, as long as the current is not higher than about 40 ma. It can be made to happen at an accurately predictable voltage, in a manner that is quite insensitive to temperature. Therefore Zener diodes can be used to ensure a constant voltage, as will be shown in the next chapter.

A neon bulb is a "gas tube diode," in which an electron avalanche takes place in a partial vacuum containing a small amount of the element neon. There is no difference between the two electrodes, so the same V-I curve is followed in both directions. Once the device has started to conduct, the voltage drop decreases. The breakdown voltage is above the 50 volt level that is considered relatively safe regarding electric shocks, so we will not trace its curve in this course.

Telephone engineers chose 48 volts as their steady dc potential (often called "50 volts") because that was thought to be unlikely to kill anyone who might accidentally touch uninsulated terminals. Nowadays it is known that some people can get hurt at lower voltages, if the electricity goes from hand to hand, across the heart, especially in humid weather, although such accidents are rare. (For "ringing" to announce an incoming telephone call, short bursts of about 20 Hz ac go through a capacitor to a bell. These can be somewhat higher than 50 volts.)

Another type of avalanche diode is the "diac." This also has a characteristic curve that is the same in both directions, because it is a symmetrical design with a PNP structure. It can be thought of as two PN diodes, back to back. Each one is designed to be able to avalanche at about 1 ma of current without being damaged. Breakdown can be as low as about 6 volts, and this goes lower after the avalanche has started, even more so than in the neon bulb. We will make use of this device in Chapter 21. As an optional experiment, it can be inserted into the curve tracer to give a curve like that in Fig. 14.4.

Rectification

The circuit of Fig. 14.5 should be assembled, but without the dashed line wires being connected. If the voltmeter is connected differently than shown, being attached *before* the diode and resistor (directly across the black center tap and one yellow secondary wire), the ACV setting on the meter will show about 6 volts, but the DCV setting will not show any voltage. Then, hooking up the meter *as shown* in the figure, the ACV setting will not show anything, but DCV will indicate about 9 volts (to be explained at the bottom of the next page).

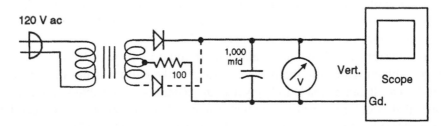

Fig. 14.5 A half-wave rectifier, becoming full-wave when the dashed line wires are attached.

The oscilloscope can be attached to the transformer secondary *before* the diode and resistor, as was done at first with the meter, and the display will be similar to the one marked "original" in Fig. 14.6. (This is best done while temporarily disconnecting the transformer.) Note that the sine wave might show some "distortion" because of nonlinearity in the transformer's iron core behavior. Also, there might be large capacitances and/or inductances already attached to the 120 VAC wires in some other room (fluorescent light inductors, air conditioning motor "starting capacitors," etc.), which can distort the waveform.

If the scope is then attached as in the figure, but the capacitor is disconnected, the display will look like the middle one in Fig. 14.6. This is because the top half of the ac wave goes through the diode, but the bottom half does not. When the capacitor is then attached, the waveform is smoothed somewhat, although residues of the "ripple" might still be visible, especially if a 1K resistor (not shown in the figure) is attached across the capacitor, to "load it down." These ripples are undesirable in a radio, because they would be heard as audible humming sound. They are undesirable in a computer, because they could be interpreted as false digital signals (to be explained in Chapter 19).

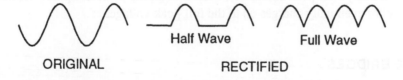

ORIGINAL   Half Wave   Full Wave

RECTIFIED

Fig. 14.6 Oscilloscope patterns with the circuit of Fig. 14.5.

With the capacitor disconnected again, the scope will show the "full wave" pattern in the figure, when the second diode is connected via the dashed line wires. This "inverts" negative half-waves and then inserts them as positive halves, fitted into the otherwise-empty time slots between the first half-waves. With the capacitor attached again, there is less ripple, even with a 1K resistor load. Also, twice as much power can be taken from this "dc power supply" circuit, without draining the capacitor to the point where the output voltage is much lower than desired. Therefore a full wave rectifier yields higher power with less ripple.

Peak Versus RMS Voltage

Looking back at the bottom of page 154, why is the voltage 9 instead of 6? The answer is related to the fact that the *maximum* part of the sine wave in the wall socket is really 170 volts, instead of 120. In other words, the true "peak voltage" is 120 times the square root of 2, which is about 170 volts. Then why is it called "120 volts?" Because a sine wave with a 170 V *maximum* ("peak") has an "effective" value of 170 divided by the square root of 2, or 120 V, so we call it "120 volts." (The square root of 2 is 1.414, and 1/(1.414) is 0.707, which are numbers that are used a lot in electronics.)

It turns out that to determine the *effective* value of a sine wave, you have to do the following. First divide up *the top half* of the ac into very small slices (the "increments" of calculus), and then *square* each one of these short-term voltages. The next step is to figure out the average value (the "*mean*") of these squared slices. Then take the square *root* of this mean. This is called the "root mean square" voltage, or "RMS." It is not the same as the simple average, and also it does not work for odd shapes such as sawtooth waves.

This 120 volts *RMS of ac*, pushing sine wave current through a certain resistance, produces the same amount of heat power that a *steady dc* 120 volts would produce, going through that same resistance. The peak voltage would give a falsely high estimate, because it only exists for very short times, so it is not used for these purposes. We therefore call the wall socket electricity "120V" (or "240V" in many European countries), even though the peak is higher. However, the capacitor eventually charges up to the full peak voltage.

DIODE BRIDGES

Another way to get "full wave rectification" is shown in Fig. 14.7. It does not require a center tapped transformer, but it does use two more diodes. It gives a higher output voltage, because the negative output wire does not have to come from a center tap. (The output is 12 times 1.414, or 17 volts.)

This circuit works as follows. During the "top" half of the cycle, "positive current" flows* through diode A in the figure, and into the capacitor, so the top half of the capacitor becomes positively charged. At the same time, negative current from the "bottom" of the secondary coil goes up through diode C, to the negative output wire. During the other half-cycle, the current coming out of top of the secondary coil is negative, so it goes down through diode B, again into the negative output.

---

* The concept of "positive current," although convenient, is really just imaginary. There is actually a negatively charged current of electrons going in the opposite direction.

What comes out of the bottom of the coil is now positive, so it goes through diode D, further charging the positive side of the capacitor. The inductor and capacitor together make up a good low pass, LC filter (see index), so the output current is relatively free of ripples.

OPTIONAL

The circuit of Fig. 14.7 (and the ones of Figs. 14.8 and 14.9) could be assembled by the experimenter as an optional exercise, but this would require sharing extra components with another experimenter or another laboratory group.

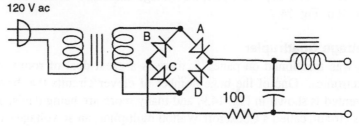

Fig. 14.7 A full wave bridge rectifier in a 17 volt dc power supply.

VOLTAGE MULTIPLIERS *(FOR READING, NOT EXPERIMENTING)*

Full Wave Voltage Doubler

The full wave power supply in Fig. 14.8 delivers 34 volts, twice the amount provided by the previous circuit. When the top of the coil is positive, the +17

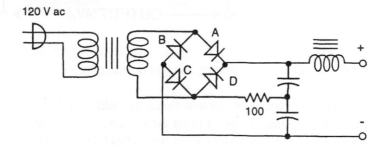

Fig. 14.8 A full wave rectifying voltage doubler.

volts goes through diode A to charge the top capacitor, as drawn in the diagram. At the same time, the bottom of the coil, as drawn, sends -17 volts to the bottom of the *other* capacitor. When the transformer is putting out zero volts (sometimes called the "crossover" or "zero crossing" point in time), both capacitors are in series, so the power supply's output terminals (small white circles) have 34 volts across them. The diodes prevent this voltage from going back into the transformer while the latter is at zero voltage. During the other half of the 60 Hz cycle, -17 volts goes through B, and +17 volts through D, to continue charging the capacitor pair in the same direction as before. Of course, when current is drawn from the capacitor pair, the voltage decreases faster than it would with the circuit of Fig. 14.7.

Voltage Quadrupler

As was mentioned on page 75, there is a great deal of room for creativity in electronics. One of the huge number of clever circuits that have already been invented is shown in Fig. 14.9, and many more are being devised every day. In this circuit, called a Cockcroft-Walton multiplier, an ac voltage can be multiplied by four.

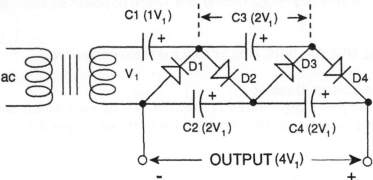

Fig. 14.9 A half wave rectifying voltage quadrupler.

If another two capacitors and two diodes are added to the circuit in the same manner, the total output can be six times the voltage, and one more set would give eight times the voltage, etc. These ideas were used by scientists named Cockcroft and Walton to build million-volt power supplies for nuclear "atom smashers," and now they are used in thousand-volt supplies for portable TV sets.

OPTIONAL READING

OPTIONAL READING

The operation of the circuit in Fig. 14.9 proceeds* as follows. When the *bottom* of the transformer secondary wire is positive, current flows through diode D1, charging capacitor C1 to the transformer voltage. When the *top* wire then becomes positive, the voltage going through C1 and also D2 is the transformer voltage ("V") *in series with* the 1V of C1, so it is 2V, and that is the voltage that C2 gets charged up to. The next time the bottom wire is positive, that 2V is put in series with the 1V of the transformer, which would have charged C3 to 1V + 2V = 3V, except that the 1V of C1 is *subtracted* from it (like batteries in series, but with one at the wrong polarity). Therefore the charge on C3 is only 2V. The next time the top wire is positive, the transformer voltage plus C1 plus C3 gets put across C4, but C2 is subtracted from it, so the voltage on C4 is 1V + 1V + 2V - 2V = 2V. However, we can take output from the total of C2 and C4, so we can get 4V out. It takes at least 4 cycles to recharge C4, so we can not take many coulombs of charge out, in any given time period, and the output current at full voltage is therefore low.

DIODE SNUBBERS

On pages 110 and 138, capacitors were used as snubbers, to absorb the inductive kick from a quickly turned-off inductor. Another way to do this is to attach a PN diode, not across the switch as on page 138, but across the coil, as in Fig. 14.10.

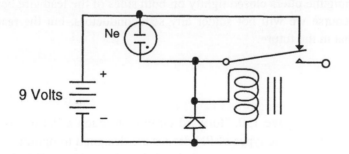

Fig. 14.10 Using a PN diode as a snubber for inductive kicks.

---

* *ARRL Handbook for Radio Amateurs,* edited by R. D. Straw (American Radio Relay League, 1998) page11.9. The publisher is in Newington, CT 06111 (www.arrl.org).

When the switch (normally-closed relay contact) is opened, the "positive" charges tend to continue downward through the coil. They would make a potentially destructive spark across the contacts, and they would light up a neon bulb if it was attached, continuing in the original upward direction through the battery, as shown in the diagram and then downward through the contacts. However, with a diode attached across the coil, these charges can easily continue upward through that diode, instead of going through the battery and contacts. The bulb will not light if the diode is present, which demonstrates the effect of the diode snubber. (This could be done as an optional experiment, if the relay and neon bulb are available.)

SOLDERING SEMICONDUCTORS

When soldering the leadwires that are attached to a semiconductor device such as a diode, the PN junction must be protected from excessive heat, in order to prevent the dopants from diffusing further than was intended by the manufacturer. (This warning also applies to the transistors discussed in the next chapter, and also to other heat-sensitive devices, some of which are shown in Fig. 14.11.) A long-nosed pliers should be clamped onto the leadwire, between the semiconductor and the solder joint, to serve as a "heatsink" and absorb any excess heat before it can get to the PN junction. The pliers can be self-clamping, if a strong rubber band is stretched around the handles, substituting for a person's hand and thus keeping the pliers closed tightly on both sides of the leadwire being soldered. In this course we will not solder any semiconductors, but the reader might have to do that in the future.

OTHER DIODES

As mentioned at the top of page 150, "forward current" through a PN diode can produce photons of light. Some types of PN diodes are designed to optimize this effect, either for visible light emission or invisible infrared emission. This device is called a "light emitting diode," or "LED," and a commonly used symbol for it is shown at the upper left of Fig. 14.11. In the middle of page 150, it was mentioned that energy can flow the other way, from photons to electrons, and devices that are designed to optimize this are called "photodiodes," as also shown in the figure. (They are different from "photoconductors" — see index.) Also, some types of transistors, not covered in this course, are light sensitive.

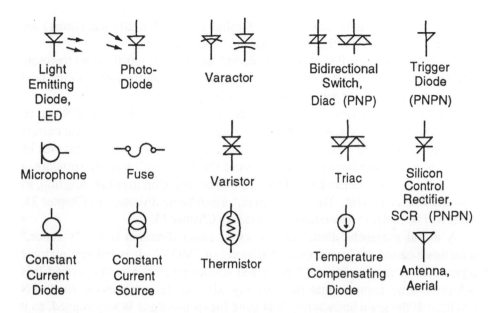

Light Emitting Diode, LED

Photo-Diode

Varactor

Bidirectional Switch, Diac (PNP)

Trigger Diode (PNPN)

Microphone

Fuse

Varistor

Triac

Silicon Control Rectifier, SCR (PNPN)

Constant Current Diode

Constant Current Source

Thermistor

Temperature Compensating Diode

Antenna, Aerial

Fig. 14.11 Symbols for other types of devices (not all semiconductors).

In an ordinary rectifying PN diode, the depletion zone is a thin insulator, so the device is really a capacitor when it is reverse biased (Fig. 14.2 on page 151). The *higher* the reverse voltage, the wider the insulating ("dielectric") layer, so the *lower* the capacitance is. This is used to make variable capacitors, for tuning radios and TV sets. A battery plus a potentiometer are wired to put a small dc voltage across a specially optimized PN diode, and this "varactor" or "tuning diode" is attached to an inductor, making a resonant circuit. The resonant frequency depends of course on the capacitance (page 126), so the dc voltage can easily be used to set the frequency. (Note that several alternative names, and also several alternative symbols, can designate many of these diodes.)

The Zener diode, a purposely avalanched reverse biased device, was discussed on page 153. A diac, sometimes called a bidirectional or bilateral switch, was mentioned on page 154. For continuity, the diac is repeated in Fig. 14.11, so it can be compared with some other devices. It also goes into avalanche, but it can do so equally in either direction (unlike a Zener). When it "breaks down" (starts avalanching), its voltage does not stay constant like a Zener's does, but instead it decreases, even more than a neon bulb voltage does (page 153 again). For this reason it is useful as a "gate" or "threshold." Once a voltage across it has built up to a certain value, current can then pour through it without as much voltage drop as a Zener or neon bulb would have. Similar devices (usually transistors) are quite valuable in constructing "logic modules" for computers.

Another device, the PNPN "trigger diode," has an extra junction and therefore is not symmetrical. More will be explained about this configuration in the next chapter, where threshold gate circuits are discussed.

If another wire is attached to the PNPN diode, a triode called an SCR can be made. This is a type of gate which can be triggered (turned on) by a small current sent into it via a third wire (shown at an angle in the figure). Again, this will be explained further in Chapter 21. If still more things are built into the structure, a very useful gate device can be made which can be turned on in *either* direction, so it works with ac current. This is the "triac," again being discussed in Chapter 21, but it will be easier to understand after reading Chapter 15.

A useful avalanche diode that works in either direction is the "varistor," sometimes called the "metal oxide varistor," or "MOV." It is not just a single crystal device as the previously described semiconductors are, but instead it is polycrystalline ceramic material, usually zinc oxide, with many small PN junctions at the grain boundaries. It is quite inexpensive and is very rugged, so it can be found as an important part inside most "surge protector power strips" for connecting multiple computer modems, printers, etc., to protect them against inductive kicks and even most "spikes" ("transients") from lightning .

A "fuse," as most readers know, protects against too much current flow by melting and thus breaking contact to the power source. It must be replaced after doing its protective job, while a "circuit breaker" (see index) can be reset after it has been "tripped" (operated to break contact), without replacing it. Sometimes a circuit breaker is an electromagnetic relay, and sometimes an SCR or Triac.

A "thermistor" is a ceramic resistor which is designed to purposely change resistance a lot when the temperature changes, while most ordinary resistors are designed to stay as constant as practical. (*Thin* film metal resistors can be extremely constant, while the cheaper carbon composition types change more with temperature. *Thick* film cermet resistors — *cer*amic plus *met*al — are in between those two, both in constancy and in price.) Thermistors are often put into circuits so their changes are exactly opposite those of the other components, and thus they "compensate" and make the resistance of the whole circuit remain fairly constant, as the temperature changes.

A "temperature compensating diode" has only two wires coming out of it, but it usually has a complex integrated circuit inside its container ("package"). Its properties self-adjust in a manner that compensates for changes in room temperature, where metal-pair thermocouples are joined to ordinary copper wires inside furnace controllers (but not in the furnace itself), so the thermocouple-to-copper junction properties are exactly matched. Thus, changes in room temperature do not affect the thermocouple reading. (This and some of the other symbols are included here because it is difficult to find them explained anywhere else. Often they are used in electronics diagrams without any explanation. An example is the radio "antenna" or "aerial" symbol, which is usually assumed to be understandable to new students, but sometimes is mysterious.)

A "constant current diode" (lower left of Fig. 14.11) also has two exterior wires but is more complex inside. It changes resistance according to the current passing through it, within reasonable limits, which tends to keep the current constant. (More in Chapter 18.) It is used to make sawtooth waves (see index) with straight ramps, instead of the otherwise exponential voltage buildups.

A microphone has a similar symbol but is unrelated, of course. Microphones can be electromagnetic like loudspeakers, or they can be variable capacitors ("electret type"), or "piezoelectric."

Piezoelectric Diodes

A piezoelectric device will be used in the next chapter. There is no standard symbol, but the lower right-hand diagram on page 91 is sometimes used, without the plus and minus signs. It is a capacitor that uses special material as the dielectric between the metal plate electrodes. This material is most often either single crystal quartz (silicon dioxide) or polycrystalline ceramic ("PZT," lead zirconate titanate). Newer types are sometimes single crystal lithium niobate. When it is squeezed, an electric voltage is generated in the capacitor. On the other hand, if an exterior voltage is applied to it, the material will expand or contract, depending on the polarity.

This type of material can be used in a microphone or loudspeaker (which we will do in the next chapter), and it can also be made to resonate mechanically at a single frequency, like a bell or a guitar string. The resonating property is quite valuable, especially because quartz can be cut in a certain direction (usually the AT plane) so that its resonant frequency is extremely constant over a wide temperature range. The frequency control is so accurate that it is used to time battery driven wrist watches and tune TV sets, from Alaska to Florida. Lithium niobate is more expensive but has some better characteristics, so it is being used to tune cellular phones and palmtop wireless computer terminals, with a special vibrational mode called SAW (surface acoustic waves). This is an example of things that were once rare and expensive but are now becoming everyday necessities for modern business operations.

EQUIPMENT NOTES

Components For This Chapter	Radio Shack Catalog Number
Clip leads, 14 inch, one set.	278-1156C
Transformer, 120V to 12V, 450 ma, center tapped	273-1365A
Resistor, 100 ohm, 10 or 1 watt	271-135 or 271-152A
Diode, silicon, 1N914, glass package *(two)*	276-1620 or 276-1122
Oscilloscope, Model OS-5020, LG Precision Co., Ltd., of Seoul, Korea	
Diode, Zener, 6V, 1N4735, 1W	2760561 or 276-565
Capacitor, 1000 mfd	272-1019 or 272-1032
Multimeter (or "multitester")	22-218

OPTIONAL

Relay, 12V dc coil	275-218C or 275-248
Neon test light, 90 volts	22-102 or 272-707
Diac ("bidirectional switch"), 6 volts	900-3156 or 900 3157*

---

* This and similar items can be purchased from the Radio Shack mail order subsidiary, RadioShack.com, P. O. Box 1981, Fort Worth, TX 76101-1981, phone 1(800) 442-7221 ,
e-mail commsales@radioshack.com,
website www.radioshack.com.

CHAPTER 15

The Bipolar Transistor

WHAT IT IS

The invention of the transistor in 1948, by Nobel Prize winners Bardeen, Brattain, and Shockley of Bell Telephone Laboratories, has greatly changed the industrialized world. The transistor can increase the "amplitude" (voltage or current) of signals such as voice messages or machine-generated numbers, without wasting much energy as heat, or taking up much space, or costing much money. This means that a signal can be sent around the world, or through an extremely complex computer, without becoming weaker, because any decrease in amplitude can be made up for by a multitude of small, cheap transistors. If the signal becomes too strong, it can easily be "attenuated" (made weaker) by a resistor, which can bring it back to the original amplitude.

Another important factor, however, is the fairly new development of "digital" technology (which will be discussed in a later chapter). This prevents "noise" (random errors) from creeping into the signal. The two innovations, transistorization and digitization, have made it practical for us to make electronic systems of almost unlimited complexity. As an everyday example, the extremely complex little microprocessors in most new automobiles have improved gas mileage, lowered pollution, and improved the effectiveness of brakes. Of course, the reader knows of many other examples, such as cellular phones, etc.

An analog of the transistor is a water faucet, where a small amount of energy can open or close a valve, thus controlling a high pressure or flow rate. In the electrical case, a small voltage or current can control large amounts of either, thus making any kind of signal much stronger. Compared to a relay, a transistor is much *faster*, and it can be *partially* "on" or "off."

The "bipolar" type of transistor is usually made from the element silicon, which has been doped so it has two junctions, instead of the single junction that was described in the previous chapter on diodes. Bipolar transistors can be of either the PNP or NPN type.

The PNP type is shown diagrammatically in Figure 15.1. A voltage across the *right-hand* PN junction is not able to pass any current, because the holes (little open circles in the diagram) are attracted by the negative charge of the right-hand battery, leaving the same "depletion zone" (often called "depletion region") that exists in reverse-biased diodes. Since there is no current flowing through the load, the transistor is in the "off" condition.

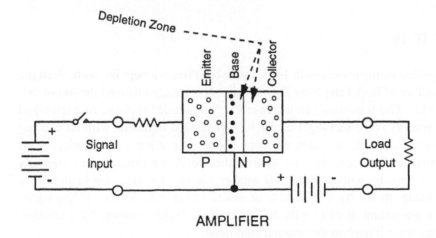

Fig. 15.1 A PNP transistor in a "common base" arrangement. (All the currents go through the same base wire.)

There is another complete circuit at the *left-hand* side of the diagram, and this is where the "signal" comes into the transistor. Now imagine that the switch has become closed. Since positive charges go to the P-type emitter, and negative charges to the N-type base, current will go through the left-hand junction. The emitter is made with intense doping, so a lot of hole current flows, but the base is purposely made only lightly doped, so there is *not much electron flow in the base.*

At the right-hand side, the collector has a strong negative charge on it, and those *extra* holes from the emitter are attracted toward the collector. The base is very thin, so a lot of the holes go across the depletion zone, eliminating it and thus allowing the right hand diode to "turn on." Only a small signal voltage is needed to control a large output voltage, so there is "amplification."

EXPERIMENTS

Common Base

Figure 15.1 showed the "common base" type of transistor hookup, in which two circuits share the same base lead. That particular variation required two batteries, each attached with opposite polarities, and this is a disadvantage of the circuit. Other variations of the common base circuit only need one battery, but they are more complex in other ways.

In the present experiment, one of the batteries is contained in the ohmmeter. (It should be noted, as was mentioned on page 25, that the red wire of many ohmmeters puts out a negative voltage.) The inexpensive transistors used in this course might be inside black plastic "packages" of the TO-92 configuration. If so, the three wires are as shown in Fig. 15.2. The leadwires should be bent *far* away from each other, starting the bends about one millimeter from the package.

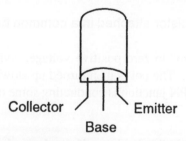

Fig. 15.2 The TO-92 plastic package for a transistor.

The first thing to do is test the transistor. (Colored marking pens can be used on the wires to distinguish PNPs from NPNs.) Using the ohmmeter, and knowing that the red probe has a negative output, touch it (or attach it via a clip lead) to the N-type base of the PNP transistor, while the black probe is touched to the collector. There should be conductivity (a fairly low resistance). Then move the red probe to the emitter, and the same thing should occur. Reversing the leads, no measurable conductivity should occur, because both PN junctions will be reverse-biased. If these are *not* the observations, then the transistor is defective, and it must be replaced.

   The wiring diagram for the common base experiment is Fig. 15.3. Note that the symbol for the PNP transistor only has one arrow, even though it really contains two diodes. The arrow is always on the "emitter." (Sometimes a circle is drawn around the whole transistor, although that type of symbol will not be used here.) Before the 9 volt battery is attached, the *ohmmeter* should show essentially infinite resistance, just as it did in the test. Thus there is no "output" current flowing.

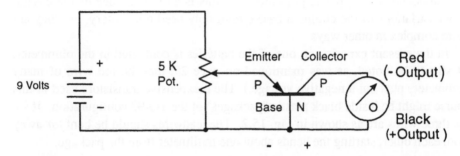

Fig. 15.3   The PNP transistor attached in a common base configuration.

   The pot is turned "down" to zero positive voltage, and its "top" terminal is then attached to the battery. The pot is then turned up slowly, and the ohmmeter will begin to show that the PN junction is conducting some output current.

Common Emitter
The *NPN* transistor is used in this hookup. If the device is in a TO-92 package, the leads will be as shown in Fig. 15.2, the same as with the PNP. Again, the first thing to do is test it with an ohmmeter, but the results should be the opposite of what was observed with the PNP transistor. That is, when the positive output of the meter is attached to the base of the NPN, and negative to either the collector or the emitter, there should be conductivity, but not when the leads are reversed.
   The circuit will be as shown in Fig. 15.4 (on the next page). An advantage of this hookup is the fact that only one battery is needed, so the common emitter is used very often. Another advantage is that either voltage or current can be amplified. In this case, a meter can measure the input current.
   The voltage that is reverse-biasing the collector is called V_{CC}, meaning the collector voltage at "cutoff" (no collector current). In this case it is about 9 volts,

since no current is flowing through the bulb. Another abbreviation is V_{CEO}, meaning the collector-to-emitter voltage, with no current flowing.

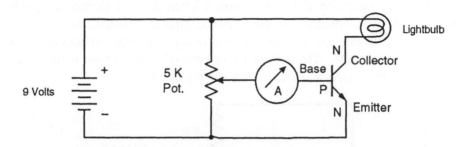

Fig. 15.4 The NPN transistor attached with a common emitter. (All the currents go through the same emitter wire.)

COMMENT

Do not turn the potentiometer all the way up to the maximum positive output. Too much current would then flow through a very small area of the pot, probably burning it out, and also the transistor might burn out. In fact, a reasonable safety measure would be to put a 100 ohm resistor in series with the base of the transistor, and the instructor might suggest that this be done.

The experiment consists of turning the pot down toward the negative lead (all the way *down* will not hurt anything), then attaching the positive battery wire to the pot, and then slowly turning up the pot until the bulb lights visibly, but stopping before the light becomes very bright.

The ammeter should show about 2 ma. From the experiments described on pages 24 and 84, it could be seen that the bulb lights at approximately 40 ma. Therefore the "amplification" is about 20, in this situation. However, this varies according to the particular configuration being used for measurement.

OPTIONAL

Replacing the meter with a direct wire, and then using it as a portable *voltmeter*, the base-to-emitter input voltage can be compared to the output voltage across the bulb. Again, the bulb should not be lit to full brightness. The input voltage will probably be too small for precise measurement, but the amplification is probably a factor of 100, approximately.

Figure 15.5 shows another way to use the common emitter, in which voltage going to Output 2 is *decreased* when the input is increased, so the signal is said to be "inverted." Inversion of the waveform is often done in electronics, partly because it is very convenient to use common emitter circuitry, with just one power source. In this hookup, the emitter is really "common" to both the input and the output circuits. The bottom 1K resistor (shown with the parentheses) does not have to be used, and a direct wire can replace it.

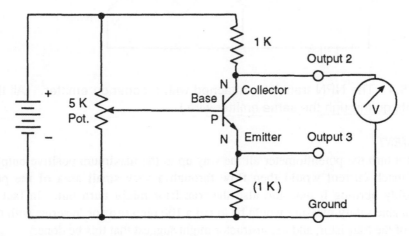

Fig. 15.5 Two alternative ways to use the NPN with a common emitter.

If the Output 3 voltage is compared to "ground," and the bottom 1K is used, then the circuit is called an "emitter follower." Increasing the input will increase the output,* so this arrangement does not invert the signal. This circuit can be a useful thing to know about when low output resistance is needed — in that case, the top 1K is not necessary, and the arrangement can also be called a *common collector* circuit. An example will be described in the next chapter, as a type of "buffer amplifier" (page 180).

Regulated Power Supply

In the previous chapter, various methods of rectifying ac to become fairly smooth dc were described. However, the rectifiers studied in that chapter would deliver voltages that are too low, if the wall socket 120 volt ac were to drop somewhat in

---

* The reader is reminded of "output" into Glass 2 of Example A on page 36, and "output" into Meter 3 in Fig. 4.2 on page 38, which are similar to the emitter follower's "Output 3."

voltage due to heavy usage from air conditioners, etc., or if there is a low resistance load attached to the dc output. What is desired, to make the TV picture be a constant size, and the computer operate reproducibly, is a dc power supply that is "regulated." That is, it can internally compensate for the ordinarily expected variations of the input ac, or the ordinarily expected variations of the load resistance. Many dc power supplies are indeed regulated this way, and a common circuit is shown in Fig. 15.6.

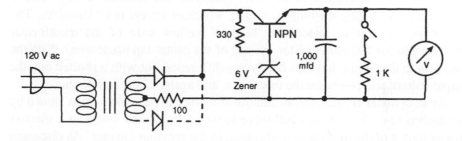

Fig. 15.6 Regulated power supply, similar to the "Output 3" in Fig. 15.5.

When constructing this circuit, the experimenter should use the 120V to 12V transformer, with a power cord soldered on, and insulated with tape. The 12 volt secondary is "center-tapped" with a thin black wire, and the 100 ohm resistor is attached to this, so the initial turn-on current does not rush into the empty capacitor too fast. (This is called "in-rush" current, and it can damage some components. It can also occur when heated filaments or electric motors are first turned on.)

Initially, the "switch is closed," (really just a clip lead attached), applying a 1K load. When plugging into the 120V wall socket, just one diode is used, and half-wave rectified dc fills the capacitor, as was done in the previous chapter. The meter reads about +5.5 volts.

The Zener diode plus the 330 ohm resistor are a potentiometer, and they apply +5.5 volts to the base of the transistor. If the capacitor voltage gets *less* positive than this regulated value, then positive charges will flow down through the 330, then up through the base and into that capacitor. When the capacitor fills to 5.5 volts or a very small amount higher, the filling process stops, until the capacitor is drained again. Thus the output voltage is maintained at very close to the desired value.

Connecting or disconnecting the 1K load makes very little difference to the output voltage. Now attach a clip lead from the collector to the emitter, "shorting" the transistor. (The unregulated voltage will now go up.) Repeat the removal and re-attachment of the 1K. This time the load has a large effect, considerably changing the output voltage, which of course is undesirable. The next experiment is to attach the "bottom" yellow wire of the transformer secondary to the 100 ohm resistor, instead of the center-tap black wire. With the transistor in the circuit, it makes hardly any difference, but with it shorted out, the output voltage will rise when the input rises, and again, this is undesirable.

As an optional experiment, an additional diode can be hooked up as shown by the dashed line. This makes full wave rectified dc, which works in a manner similar to that of the four-diode bridge used in the previous chapter. As discussed in Chapter 11, filters can do further smoothing of the output. The combination of a bridge, the transistor, and an LC filter can provide good enough dc for radios and computers. Even better results can be obtained with feedback, and that will briefly be covered in Chapter 23.

Gain

The *ratio* of the input signal (either current or voltage) to the output signal is the "amplification factor," sometimes called "gain" or "transfer function." There are several ways to specify this, depending on whether the emitter or base or collector are "common." For common emitter circuits, the gain is also called the "hybrid forward emitter" ratio, or "beta," which is:

$$h_{fe} = \beta = \frac{\text{collector current}}{\text{base current}} \qquad (15.1)$$

The h_{fe} is the usual way that transistors are labeled on their packages. Typical values for inexpensive devices are approximately 50 to 200. (See also the curves in the Appendix.)

Nonlinearity

The "characteristic curve" of input versus output, either by current or voltage (but somewhat different for each) is generally of the "s-shape" type, as in Fig. 15.7.

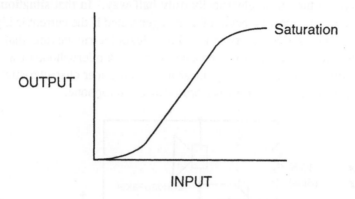

Fig. 15.7 The characteristic curve of a transistor.

Usually, a linear relationship is desired. At the top, the input is obviously not linearly related to the output, and the available output reaches a maximum at the "saturation" level. It is undesirable to work in that part of the curve with "analog" circuits, but sometimes it is useful for "digital" circuits (more in Chapter 19).

Bias

At the bottom of the curve, the input-versus-output currents are also nonlinear, and in fact it is sometimes a problem that a lot of input has to be applied before there is any useful output at all. To prevent that, a certain small amount of artificial input current is often set up to flow at all times, which puts the operation up on the linear part of the curve, ready for the first real input from external sources. This artificial input is called "bias," whether it is current or voltage.

If the signal is an alternating current sine wave, being at either of the nonlinear ends of the characteristic curve will cause "distortion" of the waveform shape. This causes the formation of "harmonics" of the original frequency, so the linearizing effect of a bias is quite important in such low-distortion applications as audio amplifiers.

Darlington Pair with Piezo Sensor

One transistor can feed its amplified signal directly to another, and this is called the Darlington circuit, named after its innovator. If each transistor has a gain of about 100, the overall gain is about 10,000. They can be bought in a single package, and a solid metal part of this package often has a hole in it, for close attachment to an aluminum metal chassis by means of a fairly large bolt. This provides a good path for heat to be conducted away, during the time that the transistors might be turned on electrically only half-way. In that situation, the device is a resistor, and a large amount of heat is generated if the current is high.

Figure 15.8 shows the experiment. The piezoelectric device that was explained in the previous chapter can be used as either a microphone or a small loudspeaker. Either way, it is not very linear in its transducer characteristics, but it generates a large signal when used as a brute-force microphone.

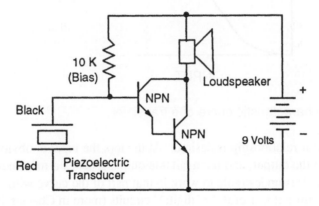

Fig. 15.8 A Darlington circuit, driving a loudspeaker.

The inexpensive piezo devices used in this course were not designed for use as microphones, and they work best if the black leads are made positive. A small loudspeaker (electromagnetic, with a paper cone) can serve as the other transducer, converting an electric signal back into mechanical sound waves. However, if it is not available, another piezo can be used in its place.

If the piezo device is held as far away from the loudspeaker as the clip leads will allow (and possibly dangling over the edge of the lab bench), it can be sharply rapped with a pen, near its edge, and a *much* louder noise will come from the speaker. However, disconnecting the 10K, and then *very lightly* rapping the piezo will not cause any output at all from the speaker. The reason is that the small signal will be at the bottom of the s-shaped curve in Fig. 15.7, and there is even *less* output than input. Therefore a bias current is introduced, through the 10K, which puts the operating point up on the s-shaped curve, where there is more gain (slope), and it is also more linear. (Some Darlingtons also have internal bias resistors.)

With a better microphone and a longer wire, a simple intercom or telephone can be made with the Darlington circuit, and the reader might want to do some optional experiments with this idea. The loudspeaker can be moved to the transmitting end, because it can act as a more sensitive microphone than the piezoelectric transducer, and if so, a 4.7 microfarad electrolytic capacitor should be put in series with this speaker, to prevent bias current from going through it. The piezoelectric device can then become an earphone at the receiving end. (However, it would be better to use a second loudspeaker at the receiving end.) A long 120 volt extension cord can be used as the connecting wire, snaking under a closed door to another room. The probes from the multimeter can be poked into the female socket at one end of the cord, in order to establish contact at that end. A portable radio can be used as a continuous sound source.

EQUIPMENT NOTES

Components For This Chapter	Radio Shack Catalog Number
Battery, nine volts	23-653
Clip leads, 14 inch, one set.	278-1156 or 278-1157
Lamp bulb, tungsten, 12 volt dc, green or blue	272-337A or 272-337C
Multimeter (or "multitester")	22-218 or 22-221
Resistors, 330 ohms , 1K, 10K (1/2 or 1/4 watt)	271-306 or 271-308
Resistor, Power, 100 ohms (1 or 1/2 watt)	271-152A or 271-308
Potentiometer, 5K	271-1714 or 271-1720
Transformer, 120V to 12V, 450 ma, center tapped	273-1365A or 273-1352
Transistor, PNP, 2N3906 *	276-1604 or 276-2043
Transistor, NPN	276-1617 or 276-2041
Darlington pair, NPN	276-2068
Piezo transducer	273-073A or 273-091
Diode, silicon, 1N914 *	276-1620 or 276-1122
Diode, Zener	276-561 or 275-565
Loudspeaker	40-252 or 40-251
OPTIONAL	
Capacitor, electrolytic, 4.7 mfd	271-998
Extension cord, 120 volt	61-2748 or 61-2747

ALSO: Large items like loudspeakers might be shared among student teams. The PNP and NPN transistors can be marked blue and red (put in separate envelopes).
* Diode types are numbered 1N#, but transistor types are numbered 2N#.

CHAPTER 16

Sine Wave Oscillators

FEEDBACK AND OSCILLATION

On the top of page 137, *positive feedback* in a relay system was seen to lead to a "latched" condition, where the relay continues to *feed* operating signals *back* to itself and therefore continue in one position, until the current is somehow interrupted by some other action. It should be noted, however, that when a relay armature operates, it hits the "normally open" contact, which is a hard barrier to any further motion. It might bounce back a little bit, but it can not "overswing." Therefore, positive feedback leads to stable latching and no further motion.

On page 138, it was seen that *negative feedback* can cause "oscillation" of the relay armature up and down. However, at the top of page 136, it was pointed out that feeding back the signal to *stop* operating (that is, *negative* feedback) took several milliseconds to occur. With a carefully timed *delay* (see the text just under Fig. 12.7), the up and down motion becomes complete, flipping all the way in both directions. The delay was provided by a capacitor plus the resistance of the coil, making an RC circuit with a time constant, τ (page 95).

Transistors are so fast that their natural delay times are in the billionths of a second, so we ordinarily use capacitive delays to produce oscillations with somewhat longer time constants. (For example, to produce a musical note that we can hear, the delay should be roughly a thousandth of second.) We can use either *delayed negative* feedback (bottom of next page) or *fast positive* feedback (top of page 182). With transistor circuitry, it is more often directly *positive*.

177

Phase Shift Oscillator

Consider the type of transistor oscillator shown in Fig. 16.1. The output of the left-hand transistor amplifier is taken at the collector, just like Output 2 on page 170. However, this is really a type of "signal inversion," where an *increase* in the positive voltage going into the base turns on the transistor, causing a *decrease* in the voltage coming out (into the capacitors) at the collector terminal. (This is the meaning of "inverting" as previously described on pages 155 and 170, not the other meaning at the bottom of page 105.) Thus a sine wave going into the base would be turned upside down (inverted) when it comes out of the collector and starts going to the "network" of three capacitors and resistors.

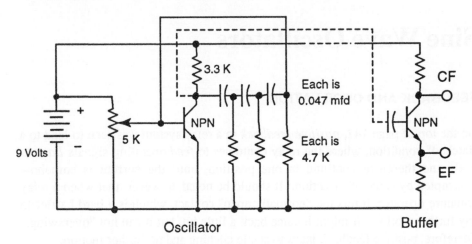

Fig. 16.1 Phase shift oscillator (with dashed lines not being connected). Buffer amplifier is for external attachments, as described on page 180.

Without the capacitor network, the *inverted* signal fed back to the base directly would give us *negative feedback.* However, we will now *delay* it 180°, which is called a "phase shift." Then we will have *delayed negative feedback.*

For a sine wave, being delayed 180 degrees is the same as being inverted (see picture on page 98). Thus, the network of capacitors and resistors inverts it *again*, turning it "rightside up." The delayed signal from the network goes up the wire as shown here, and then back into the transistor's base, making it an even stronger signal. Thus, the stronger it is, the stronger it gets, which is *positive feedback,* as the *overall* result. The RC network shown in the diagram will accomplish this at one certain frequency, but not at other frequencies. With the R and C values shown, it happens to be at about 400 cycles per second.

This self-enhancement continues until the left-hand capacitor charges to the full 9 volt battery potential, and then there is no more voltage to pump it up further. However, all wires carrying current, even straight ones, have some slight inductances, because they all generate magnetic fields. This tends to keep the current flowing a very small amount *past* the full 9 volt charge, so there is a slight "overswing," just like with a mechanical pendulum at its lowest point. (Sometimes this is called "overshoot," or "the flywheel effect.")

When the maximum voltage finally is reached, the charging stops, so the current stops. Inductance only has an effect when current is flowing, so the tendency to overswing also stops eventually, and the voltage starts back down to 9 volts. As soon as it goes down, the whole self-enhancement action (*positive feedback*) operates in reverse. That is, the more the current goes down, the more it goes down even faster, until the potential on the first capacitor reaches zero volts. But the slight inductance of the wires then carries the voltage to slightly negative values, until it finally does stop. Then the whole cycle swings back upwards again. The up and down changes are, of course, "oscillations."

Unless there is another transistor pushing things to minus 9 volts (a "push-pull" circut), which is not the case with this simplified system, the sine wave will be somewhat distorted, and other factors also distort it slightly. However, the wave shape is generally "sinusoidal," or close to a geometric sine function.

The network of three RC filters, as shown in the diagram, has to "delay" the sine wave a total of 180° in order for the circuit to start oscillating. Each RC pair delays it 60°. (This "delay line" plus the top wire are a *positive* "feedback loop.")

OPTIONAL

To determine the frequency at which that 60° delay takes place, the phasor diagram on page 101 can be referred to again. (The reader might keep a finger at that page, because it will be used once more, in a moment.) From high school geometry, it should be remembered that an equilateral triangle (all three sides the same length) has 60 degree angles at all three corners. Drawing a vertical line down the middle of the triangle creates two halves, each one of which is a 60° - 30° - 90° triangle. Looking further at one of those, if the hypotenuse is two inches long, then the shortest side is one inch long. On page 101 of this book, equation 9.6 reminds the reader that, if the hypotenuse is 2, then the sum inside the root sign must be 4. If "R" is 1 (the short side of the triangle), then "X^2" must be 3. Therefore X (the remaining side of the 60° - 30° - 90° triangle) must be 1.73 inch. Going back to the phasor diagram on page 101, if R is 1 ohm, then X_c is 1.73.

If the resistance is 4,700 ohms, as in Fig. 16.1 above, then X_c is 1.73 x 4,700. The formula for frequency versus X_c is eqn. 9.5 on page 99. Plugging 4.7×10^{-8} farads into C, the frequency will be about 400 cycles per second, which is in the audible range. If a loudspeaker could be hooked up to this oscillator while it was operating, a fairly low-pitched tone could be heard.

Another part of the circuit that needs to be explained is the 5K pot. This provides "bias" to the base of the transistor (see index, if needed for explanation). In initial experiments with oscillators, it is a good idea to use a variable bias adjustment, optimizing the amount so the current is sufficient to start the cycles.

The "buffer" amplifier is a good idea, so that output loads (resistors, etc.) do not affect the characteristics of the oscillator. For example, a low resistance such as an 8 ohm loudspeaker might drain off so much current that the circuit could not continue oscillation. Also, the inductance of a loudspeaker, added to the circuit, might affect the frequency. By attaching the dashed line wires and thus making the output signal go through a capacitor to prevent a dc load, and through a transistor to draw very little ac, any load would be "isolated" from the oscillator.

The oscillator's output is usually taken from the collector, but the buffer's output can be either from its collector (a "collector follower, CF") or its emitter (an "emitter follower, EF"). If a low impedance device such as a loudspeaker is to be driven, the *emitter follower (EF)* can be used, with *no collector resistor.* That would be a *common collector* buffer, because *both* the *oscillator* that is the "driver" of the buffer, *and* the *load* might be attached *directly* to the *same* "common" +9 volt "bus" or "rail" (see index if necessary).

An alternative is to use the CF (common emitter) output, with a low resistance, or even zero resistance in the emitter leg of the circuit. However, some resistance is usually advisable there, to prevent excessive base current.

It should be noted that the common emitter (CF) arrangement can amplify both voltage and current, but the common collector (EF) only amplifies current. The reason for the latter is that the base voltage must be higher than the emitter voltage in order for base current to flow. Therefore the emitter voltage must be low, and we can not get increased voltage. This is why common collectors are not used very often, except for buffers and a few other applications where there is a need for low output *impedance* (equal to output *resistance*, in this example).

We will not make this circuit, because it requires more components than are ordinarily used in the course. It is not "robust" — that is, it is quite easily disturbed without a good buffer. Instead, we will construct an oscillator that can stand having a loudspeaker right in the circuit, and then one that can be made with fewer capacitors.

EXPERIMENTS.

Armstrong Amplifier and Oscillators

Another way to get positive feedback is to use a transformer, as in Fig. 16.2, to send a signal from the collector, back to the base, in the proper phase for positive reinforcement. This was invented by Edwin H. Armstrong, one of the greatest electronics inventors of all time. (He developed FM radio and also invented the "superheterodyne" principle, which is used in almost all radio and TV receivers in the world.) The idea was originally applied to amplifiers, not oscillators, because it very much increases the amplification factor, due to the positive feedback. The more it amplifies, the more it amplifies further, which can lead to amplification factors of a thousand, limited only by the output capability of the transistor. (This and also negative feedback in amplifiers will be discussed further in Chapter 23.) If the amplification is high enough to overcome resistance of the wires, *continuous oscillations* will be generated, instead of just a one-time amplification of some external signal.

When assembling the circuit in Fig. 16.2, a pair of 4.7 mfd capacitors are put in the collector and emitter positions, C and E, but none is in the optional base position labeled B. The antenna (radio aerial), indicated as a triangle, is not attached initially.

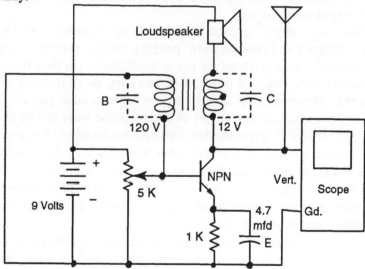

Fig. 16.2 Armstrong oscillator.

The 12 volt coil (yellow wires) of the transformer and its 4.7 mfd capacitor are a parallel resonator (page 126). A sine wave of the right frequency will continue oscillating inside that resonator for a while, until resistance "damps" it and causes it to die out, but other frequencies would die out much sooner.

The sine wave of the right frequency comes out of the 120 volt coil (heavy black wires), at about ten times higher voltage. If the coils are oriented properly (see page 111 if necessary), there will be *positive feedback* — then the waves will not be damped, but instead they will continue, once they are started by the bias pot. The 5K pot is rotated, near its middle position (neither end position should be used), until a sine wave becomes visible on the scope. Since the transformers used in this course were not designed for this application, the phase relationships are not labeled, so the experimenter will have to try reversing black-versus-black leadwires until the right orientation is obtained for producing the first oscillations.

The oscilloscope can be set up as on pages 76 and 77, starting with AUTO in order to see the line, and then switching to NORM. The TRIGGER SOURCE should be VERT. The TIME / DIV can be about 1 millisecond in the beginning, and it can be raised or lowered until several sine cycles become visible on the screen. Then the VAR knob (at upper right corner of an LG scope's control panel) is slowly rotated until a single sine wave is visible. The TRIGGER LEVEL should be made a little bit positive or negative, to stop the waves from appearing to move too fast.

The wave might be highly distorted, or it might even be a group of diminishing waves ("damped wave" pattern), because neither the transformer nor the transistor were designed for use in oscillators. (In fact, these inexpensive transistors were designed to be all the way on, or all the way off, but not in between. However, they can be used for various other purposes, such as the applications we are constructing.) A distorted sine wave can be analyzed, using the "Fourier series" approximation, into the "fundamental" frequency sine shape, plus various "harmonics" (whole-number multiples of the fundamental frequency, such as 2x, 3x, and so on). The waves generated with these simplified circuits sometimes have more harmonic content than fundamental. Sometimes what is generated is nearly a square wave, which is a fundamental sine wave plus a large number of odd-numbered harmonics.

The transformer inductance plus the parallel capacitance is called a "tank circuit." It can be attached to any of the three transistor wires, but in this case (position C), it is being used on the collector, which is the most commonly used position. (If the inductor on the base has no capacitor, it is not a tank circuit.)

The 1K emitter resistor helps to adjust the bias, and it also prevents excessive current from flowing. The 4.7 mfd at position E is sometimes called a "speedup capacitor," and it is also used with ac amplifiers, in order to let the transistor work with fast-changing ac currents, but without changing the dc bias.

The next experiment is replacement of the 4.7 mfd capacitors by two 0.1 capacitors. When the smaller capacitors are used, they will not let enough coulombs flow for the loudspeaker to make an easily-audible signal. A sine wave will be visible on the oscilloscope if the TIME / DIV is set to be faster, but probably nothing will be heard from the speaker. The frequency should be around 10 kHz or so, depending on the inductance of the particular transformer. The scope horizontal "time base" (sawtooth speed) must be faster, in order to catch a single sine wave. Again, the VAR and the TRIGGER LEVEL knobs might have to adjusted, in order to freeze the display pattern.

OPTIONAL

A piezoelectric transducer can be put across any of the capacitors, which might make a high-pitched audible sound, because it responds to higher frequencies. More will be said about piezo devices at the end of this chapter. Using the center-tap would also raise the frequency. This can be tried as another optional experiment. (The collector capacitor can be moved to position B in the diagram, and oscillations will be visible on the scope, but at a *lower* frequency, with the higher inductance of the 120 volt coil in the tank circuit.)

With 0.1 mfd across the emitter resistor, but replacing the other capacitor at position C with a 0.01 mfd capacitor, the frequency observed on the scope is in the megahertz range (about 1 microsecond cycle time). Of course, the TIME / DIV must be readjusted to much faster speeds. This is in the AM radio frequency band (500 to 1500 kHz).

As still another optional experiment, the antenna can be attached. This is any long wire, stretched out fairly straight. An easy option is to use a 120 VAC extension cord, with a clip lead attached to one of its male plug prongs. A portable radio nearby, tuned to a quiet place on the dial, will make sounds when the oscillator battery is turned on and off. The various harmonics of the distorted waveforms are mainly what are being heard, covering a wide range of frequencies. Just as it was in Chapter 1, the Armstrong-developed *FM* radio is mostly immune to such interference, and probably no sound can be heard from this experiment if an FM receiver is used.

A radio transmitter such as this oscillator would be illegal if it was more powerful, but without a good ground at the negative wire of the battery, the radio waves are too weak to penetrate out of a school building that has a steel framework (also true in the experiment of Chapter 1).

Colpitts Oscillators

The frequency of oscillation can be adjusted by the simple LC network shown in Fig. 16.3. In this case, resonating feedback goes to the emitter, causing a voltage drop across the 1K resistor, which *decreases* the current through the base (similar to SSN, page 60). However, the *negative feedback* comes from an *inverted* signal.

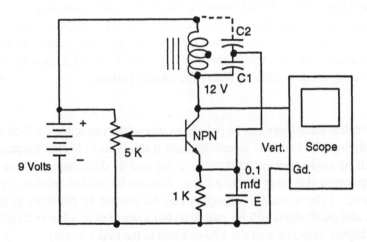

Fig. 16.3 One of many possible variations of the Colpitts oscillator.

This circuit is quite tolerant of variations, and the experiments can start by using the 12 volt coil of the transformer (yellow wires). The C1 and C2 capacitors can be 4.7 and 0.1 mfd, respectively, with the dashed wire being attached in the first Colpitts experiments. The pot is rotated (but *not* all the way to its ends) until the scope pattern shows waves. The C1 and C2 capacitors can also be 0.1 and 0.01, or both 0.01, etc. The component marked "E" is a "speedup capacitor," as described at the bottom of page 182.

If C1 is 0.01 mfd, then the dashed line does not have to be attached at all, in which case there are only two capacitors. The reason this works is that there is sufficient "distributed capacitance" inside the inductor, and between wires, so there is enough phase change to get oscillations going. (This is similar to the case of slight inductance with even a straight wire, as mentioned on page 179.)

Troubleshooters might have to search for feedback from slight inductances or capacitances, causing undesirable "parasitic" oscillations. Sometimes an internal resistance way back in the power supply gets loaded like the emitter resistor here, causing delayed fluctuations in voltage, which can feed back to a transistor and start oscillations, from either direct positive or delayed negative feedback.

OPTIONAL

Elevator controls are often strictly electronic *touch-sensitive* buttons involving no mechanical motion. In Fig. 16.3, use *the center tap,* with E = 0.1 mfd, C2 = 0.01 mfd, and the scope set on 0.5 V/DIV and 0.5 μs/DIV. *Make a capacitor* for C1 as follows: two pieces of aluminum foil about 8 X 10 inches are the conductive plates, separated by two pieces of 8.5 X 11 inch paper. Using the lowest potentiometer setting that will cause oscillation, put your hand between the two paper sheets, or actually touch the aluminum foil. The scope pattern should change. A complex filter and relay could be designed to "switch" due to this change, in order to control elevators, etc. (The experimenter might try replacing C1 with a piezo diode, to make varying sounds.)

ACCURATE TIMING

The piezoelectric transducer has some fixed frequencies for its mechanical vibrations, just like a bell or a piano string or a tuning fork. It has thin metal plates attached to each side, making it a capacitor. The symbol is at the right-hand side of page 91 (without the charges) and the left-hand side of page 174. If it is made out of quartz, cut in the correct crystallographic direction (usually AT, as mentioned on page 163), its vibrational frequency is very insensitive to temperature. For this reason, a quartz piezo "crystal" is often used as a tank circuit capacitor in radios, cellular phones, etc., to allow tuning the oscillator part of the circuit to some exact frequency.

Very slow oscillations, with long times, are not accurately obtainable in ordinary resonant circuits. For this reason, when long cycles are being timed, either an electric clock motor driving a cam-operated switch, or else an integrated circuit of the LM555 type, or some other complex device is ordinarily used, instead of a simple oscillator. There are many books about the inexpensive and versatile LM555, and it is recommended that the reader obtain one of these if more information about timing is needed.

BLOCKING OSCILLATOR

The Armstrong oscillator on page 181 can be modified to produce only a single "pulse," which would be half of a complete up-and-down sine wave. This is called a "blocking oscillator" or a "one-shot oscillator," and it would be similar to the relay-operated pulse generator on page 134, using the NC contacts and making the pulse of type B on page 135. That pulse was only 10 ms long, as described at the top of page 136, but it could have been made longer or shorter by using a different capacitor. Similarly, the Armstrong oscillator can be modified with an capacitor which charges up after a certain short time. A transistor can be used instead of the relay, to have much faster action. (Later we will make something similar, as described on page 189, where the capacitor charges up and *stops* the action after a certain length of time.)

Another way to make a blocking oscillator (which we are only going to discuss but not actually make, unless the reader wishes to "hack around" with some optional trial circuits) is to purposely reverse the yellow wires of an Armstrong oscillator to get *negative* feedback. Then suddenly attaching the battery can make a single pulse appear on the scope, but no more will occur. The size of the capacitor, in either the base tank circuit of Fig. 16.2 (B) or the collector tank (C) will control the length of that single pulse. These designs are often used for pulse generation in electronics, usually with a variety of other resistors and capacitors to fine-tune the characteristics.

EQUIPMENT NOTES

Components For This Chapter	Radio Shack Catalog Number
Battery, nine volts	23-653 or 23-553
Clip leads, 14 inch, one set.	278-1156 or 278-1157
Transistor, NPN	276-1617 or 276-2041
Potentiometer, 5K	271-1714 or 271-1720
Resistor, 1K (1/2 or 1/4 watt)	271-306 or 271-308
Transformer, 120V to 12V, 450 ma, center tapped	273-1365A
Capacitor, electrolytic, 4.7 mfd *(two)*	271-998 or similar
Capacitor, 0.1 mfd *(two)*	272-1053 or similar
Capacitor, 0.01 mfd *(two)*	272-1051 or 272-1065
Loudspeaker	40-252 or 40-251
Extension cord, 120 volt	61-2748 or 61 2747
Oscilloscope, Model OS-5020, LG Precision Co., Ltd., of Seoul, Korea	

OPTIONAL

Portable AM/FM radio	12-794 or 12-799
Piezo transducer	273-073A or 273-091
Aluminum foil and paper *(two of each)*	

CHAPTER 17

Multivibrators

WHAT THEY ARE

The word multivibrator means a circuit that can be "switched" all-the-way on and then all-the-way off, something like a relay. Some run continuously, like the relay buzzer circuit on page 138, and some wait for external signals to either turn them on or off, like the latching relay on page 136. Instead of generating sine waves as an output, multivibrators usually generate square waves like the ones shown on page 135.

EXPERIMENTS

Astable Square Wave Generator

If a transistor can be arranged to have *delayed negative* feedback, similar to the relay buzzer example on page 138, it can oscillate continuously. This is called "astable," meaning not stable. That is, it will not remain in one state, but instead it flips back and forth between "on" and "off."

With a transistor, if the feedback was *immediate*, there would be *n o* oscillation. Instead, the "negative" feedback would simply decrease the degree to which the transistor would turn on, so it would only go part-way on and then remain with that partial status. In fact, this does happen with certain other types of amplifier circuits (to be studied later) but not with multivibrators. However, if the feedback is *delayed* enough for the transistor to turn on completely *before* the negative effect takes place, then it can go all-the-way on, and *later* that negative effect will make it go all-the-way off, just like the relay did when it had capacitor delay. (By the way, it should be noted that the capacitor on page 138 could have been attached across the coil, instead of across the NC contacts, with similar effect. Often a delay can be brought about by several alternative means.)

It is possible to have *immediate positive* feedback, and also have the *delayed negative* feedback, and in fact those two actions are what we will get in the first circuit to be assembled in this chapter. The positive action makes the circuit "switch" (go *completely* on or off) very *fast*, once it starts. Because the switching is enhanced by the positive feedback, the output can be a "square wave," where the voltage goes from zero up to 9 volts quite fast (once it starts to change), hold there because of delay, and then go down to zero very fast.

In the circuit of Fig. 17.1, current goes through the low-resistance bulb, through the capacitor, and into the base of the *left-hand* transistor, turning it on.

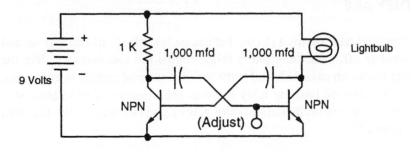

Fig. 17.1 An astable multivibrator that operates at a low frequency.

Since it is turned on completely, the + voltage on the left-hand capacitor is practically zero, so there is no current going into the base of the *right-hand* transistor, and it is turned off.

Eventually the right-hand capacitor will become *partly* charged and then not pass so much current. The left-hand transistor will then start to turn off, and its collector voltage will no longer be short-circuited to "ground" (really just the negative battery wire). Therefore current will start going through the left-hand capacitor and into the base of the right-hand transistor, turning it partly on. This will be positive feedback for the process of switching, because the more it happens, the more it will happen even further. In other words, the more the right-hand transistor turns *on*, the more that action turns *off* the *left-hand* transistor, and the more that turns *on* the *right-hand* transistor even further.

The switching transition, once the positive feedback starts to take effect, is much faster than the charging of the capacitors. Therefore the switching action can be completed, long before the negative feedback from other half of the circuit will reverse things. This is similar to the relay circuit, where the armature motion is much faster than the charging of the capacitor.

The astable multivibrator, when it is assembled as per Fig. 17.1, continues to switch back and forth. One half of the cycle might not proceed as fast as the other half, because the bulb has less resistance than the 1K resistor. If the bulb does not go out, which does happen with some transistor samples, a 10K resistor can be attached with clip leads between wire in the diagram marked "adjust" and the negative battery wire. This will put a little bit *less* voltage on the right-hand transistor base, making it turn off easier. In general, the timing of each half-cycle will probably not be equal, since the collector resistances are unequal. As an optional experiment, a voltmeter can be put across a capacitor, and the voltage buildup can be watched. Also, the capacitors can be replaced by smaller ones, and oscillations will be faster. The 4.7 mfd capacitors would make switching go too fast for the bulb to respond visibly, although an oscilloscope could easily track the voltage changes. At any rate, an output voltage taken from either collector, referenced to the negative battery wire, would be a "square wave."

Monostable Pulser
The circuit of Fig. 17.2 should be made, by replacing the 1K resistor and a capacitor. When the battery's positive wire is attached, the light goes on, but soon the capacitor charges and the light goes off again— it is just a "pulse" of light.

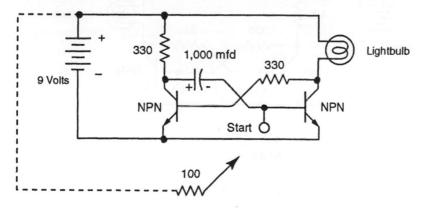

Fig. 17.2 A monostable multivibrator that generates a pulse.

To get another pulse, the battery can be disconnected and then connected again. A different way to get another pulse is to attach the dashed-line wire and its 100 ohm resistor. Then the probe (shown as an arrow, but really just a wire) is touched to the "start" terminal for two seconds. The capacitor becomes discharged, because both of its plates are connected to the same positive battery terminal. When the probe is then removed, the capacitor starts to charge again, as current flows through it and into the transistor base on the right. That transistor turns on, re-lighting the bulb, and also short-circuiting any current that might otherwise go through the 330 ohm resistor, so the other transistor is not turned on.

In a few seconds, the capacitor fills up, and the light *suddenly* goes out again, similar to the sudden turn-offs in the astable circuit. Again, an optional extension of this experiment would be to attach a voltmeter.

This circuit is referred to as being "monostable," because it is only stable in the "off" condition for long times, but not in the "on" condition. With a *small* capacitor, it is used to make *short* pulses of electricity, and this variation is called a "pulse generator." It is a useful item to have in one's "intellectual toolbox," whenever a short burst of voltage or current is needed.

Bistable Flip-Flop

The circuit of Fig. 17.3 should be constructed. It is "bistable," because it is stable for long times in either state, with the bulb off or on. It can be "set" by touching the positive probe to one of the "terminals" (really just clip leads) and "reset" to

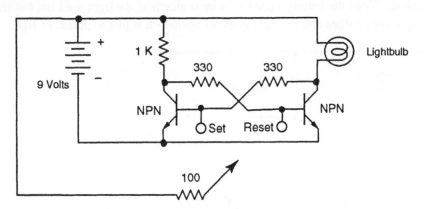

Fig. 17.3 Bistable "flip-flop" multivibrator.

the other state by touching the other one. Because it can go to either state equally well, it is called a "flip-flop," although the astable and monostable circuits are sometimes called that also. If the bulb is replaced by another 1K resistor, so that attaching the battery could *equally* lead to either transistor being turned on, this type of flip-flop is called an SR type (set/reset), or sometimes an RS type, which are the same thing.

With the bulb or another small resistor, the left-hand transistor will *preferentially* be turned on, when the battery is first attached. This type is called a JK type. (The reasons for choosing the letters J and K are obscure.)

More complex flip-flops, involving diodes, capacitors and additional resistors, can be arranged so that a single incoming wire does both setting and resetting. *Every other* input pulse *sets*, and then the *next* pulse *resets* the status. Therefore the flip-flop divides the number of incoming pulses by two. An output from one of the two transistors goes to another flip-flop, and so on, with a long string of them being connected together.

This is called a "shift register," and a pulse coming in at one end can "shift" (change the status of) the other ones down along the whole string of flip-flops, like the wind blowing a ripple along a field of wheat. A "register" has temporary memory and can "store" (remember) data. The shift action can be arranged in complex ways to do mathematical logic operations, such as additions and subtractions. More input wires can be hooked up so that some wires prevent ("inhibit") certain actions and others allow ("enable") them, depending on certain logical combinations of input pulses. Thus flip-flops are the basic building blocks of digital computers. They are used in "random access memory" ("RAM") registers, as well as in the "logic gates" that do the actual computing.

EQUIPMENT NOTES

Components For This Chapter	Radio Shack Catalog Number
Battery, nine volts	23-653
Clip leads, 14 inch, one set.	278-1156C
Lamp bulb, tungsten, 12 volt dc, green or blue	272-337A
Transistor, NPN *(two)*	276-1617 or 276-2041
Resistors, 330 ohms, 1K *(two each)*	271-306 or 271-308
Resistors, 100 ohms, 10K	271-306 or 271-308
Capacitor, electrolytic, 1,000 mfd *(two)*	271-998 or similar

OPTIONAL

Multimeter (or "multitester")	22-218
Capacitor, electrolytic, 4.7 mfd *(two)*	271-998

CHAPTER 18

FETs and Tubes

VACUUM TUBES

As mentioned earlier in this book, modern electronics would not be practical without "amplification" (see pages 165 and 166 if necessary), because this allows a large number of operations to be performed in series, without any total loss of signal strength from various resistances. The first practical amplifying device

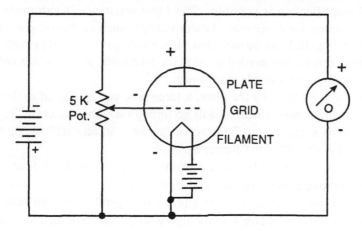

Fig. 18.1 Triode vacuum tube amplifier. (Compare to page 168.)

was the triode vacuum tube, invented by Lee deForest in 1906. The symbol for a triode is in the middle of Fig. 18.1, and it is a concise description of what is actually inside the tube. (An ohmmeter provides "plate voltage" in this diagram.)

As most readers know, the vacuum tube (sometimes called an "electron tube" or just a "tube") is a glass cylinder, closed at both ends, with some wires going through the glass, with "glass-to-metal seals." The air has been pumped out, leaving a good vacuum inside. A tungsten filament is heated to about 750°C, where it emits red light (in comparison to an incandescent lightbulb filament, heated to about 1500 °C and emitting white light).

Not only is red light given off, but electrons are also emitted, and these can be collected by a metal plate that is about a centimeter away and has a positive charge. Electrons can not go in the opposite direction (emitted from a cold plate), so a diode tube can act as a rectifier, converting ac to dc. This phenomenon was reported by Thomas A. Edison in 1883, but like a few of his other* important discoveries, Edison did not appreciate its practical possibilities, and it was mainly ignored for several decades. (John A. Fleming of England later made practical diode vacuum tubes, which he used for radio and other applications, and in fact he was knighted for his work.)

A "grid" woven from thin wires can be placed between the filament and plate, and if a negative voltage is applied, it will quite effectively repel the electron current, and prevent that from reaching the plate. About 3 volts with less than a microampere of current into the grid can cut off 300 ma at 300 V at the plate, so a great deal of amplification is possible. The input resistance is extremely high, which is an advantage for many uses. Lower voltages such as the one provided by the ohmmeter in Fig. 18.1 can be used, but higher voltages work better with tubes, and fairly high currents are needed to heat the filaments, so tubes are not very practical with pocket-size portable electronics.

Instead of being stopped by the plate, a large hole can be drilled in the plate, and many of the high speed electrons will go through it. This electron "ray" can be directed toward a transparent fluorescent "screen," in an oscilloscope or TV picture tube, as was described in Chapter 8.

There are many other types of vacuum tubes. Some have four or five electrodes, to improve power handling capabilities and for other purposes. Some have a small amount of neon or other gas, so that conduction can take place without heating a filament, although fairly high voltage is then needed to ionize the gas. Instead of the red light from a neon tube, some types give off ultraviolet light from a little bit of argon gas or mercury vapor, which can be used to ultimately produce white light through fluorescence. However, the extremely small size and low power needs of transistors have caused tubes to be used less and less in new designs. (See also the curves in the Appendix.)

---

* Another one was radio, to be discussed in the next chapter.

FIELD EFFECT TRANSISTORS

JFETs

Instead of making a transistor that conducts across both PN junctions when it is turned on ("bipolar"), a transistor can be made with just a single PN junction that participates in the action ("unijunction"). One such device is shown in Fig. 18.2. The rectangles in the middle of the diagram are solid material, with some regions that are P-type but do not conduct appreciable current, and one region that is N-type. In this case an ohmmeter provides the voltage and current, very similarly to the way it worked in the circuit on page 168. This device is *normally on,* so if the 5K pot is turned down so there is no voltage on the "gate," then "positive current" goes into the upper left corner (as shown), through the metal, down through the continuous N-type silicon, and out of the transistor through the other metal. (The "Ox." regions are silicon dioxide insulators.)

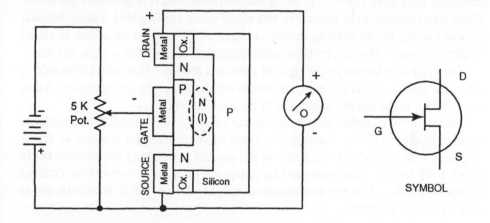

Fig. 18.2 Cross section of a JFET, with the symbol shown at the right. This device is *N-channel,* is *normally on,* and works in *depletion mode.*

That diagram is not "drawn to scale," and the rectangles shown are actually only a few microns (micrometers, or millionths of a meter) in size. The metal is usually an aluminum thin film about a micron thick, and the whole configuration is somewhat more complex than shown in this simplified diagram. The P-type silicon to the right, as drawn, is just excess material that gives mechanical support to the smaller active regions that do the conducting. It is usually referred to as the "substrate."

To *turn off* the transistor, the 5K pot setting can be raised to provide a negative voltage. (It should be noted that the battery polarity is opposite to the way it has been drawn in the previous diagrams of this book.) This negative charge on the smaller P-type region repels electrons from the N-type conductive "channel." The combination of P-type and N-type is a PN junction, and it is reverse-biased, so it does not conduct. There is a depletion zone, just as shown on page 166, so the silicon inside the dashed-line oval becomes intrinsic (I-type, as symbolized by the I in parenthesis) and it stops conducting.

This action has been brought about by the negative *electric field* from the middle metal electrode, not by carriers flowing into the region, so the whole device is called a "field effect transistor," or "FET." Since it has only one PN junction that participates in the action, it can also be called a "unijunction transistor," as mentioned above. Sometimes it is referred to as a "juncFET" or "JFET." It has extremely high input resistance, like a vacuum tube, and it is therefore useful for things like thermocouple interfaces and many other applications where minimal current loading of the driving device is required. This type of action is called "depletion mode." Overall, its behavior is quite similar to that of a vacuum tube.

The symbol is shown to the right of the cross section. One metal electrode is called the source, one is the gate, and one is the drain, similar to the emitter, base, and collector in the bipolar transistors of the previous chapter.

This is an "N-channel" device, where the current goes through N-type silicon. Another type of JFET has the opposite types of semiconductor regions, so it is a "P-channel" device, where the arrow of the symbol is touching the channel but is aimed away from it. The gate must be charged + in order to turn off the channel by repelling holes. It is not as common as the one shown, but it does exist and is used for special purposes.

Constant Current Diode

There are many applications for JFETs, and one interesting use is in making a *current* regulator (called a "constant current diode"). The total effect of this is similar to that of a *voltage* regulator, as in the diagram on page 171, except that current is regulated instead of voltage. This can be a very simple circuit, as shown in Fig. 18.3. Looking at the *negative* current that flows *upward* through the resistor, some of it will be sent to the gate, tending to turn it further off. There is negative feedback, not delayed, so that the more the current flows through the transistor, the more it turns off. Thus, less current flows, until some equilibrium is reached, at which point the current becomes constant. The JFET and potentiometer are all within the "constant current diode" symbol, which appeared previously on page 161.

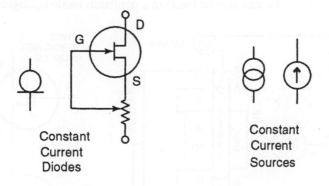

Constant
Current
Diodes

Constant
Current
Sources

Fig. 18.3 An N-channel JFET wired to be a current regulator, with the symbol shown next to it. The other two symbols (at the right-hand side) include power supplies such as batteries.

These devices are usually sold as diodes, with just two wires coming out of each package. They can be further simplified by eliminating the resistors, so they are just JFETs with the sources connected to the gates. Instead of adjusting the current, individual devices are "sorted" by measuring the actual constant current value (which varies randomly from one sample to the next) and labeling each one according to this measurement. They can also be made with bipolar transistors, but more resistors are required. Of course, the current available can not be any greater than the capability of the battery or other power supply, like an ac rectifier.

The symbol with the two overlapping circles includes the battery or other power supply. The symbol with the upward arrow is used in theoretical work (regarding the Norton theorem, for example), but not usually in ordinary circuit diagrams.

MOSFETs

A different kind of field effect transistor is shown in Fig. 18.4, the "metal-oxide-semiconductor" or "MOS" FET. Here there is insulating silicon dioxide to prevent gate current from going into the main semiconductor, not a reverse-biased junction. It is sometimes called an "IGFET," because of the *insulated gate*. It is a normally off device, which has to be *turned on* by some sort of action, and it is called an "enhancement mode" device. (The JFET described in the previous paragraph is not in enhancement mode, because the PN junction would have to be forward-biased.) IGFETs can also be made in a depletion mode configuration.

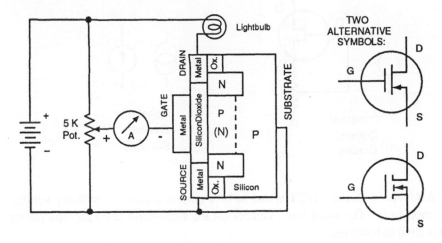

Fig. 18.4 Cross section of a MOSFET, with symbols at the right. This device is *N-channel*, is *normally off*, and works in *enhancement mode*.

In the figure, if the pot is turned to zero voltage, the battery voltage trying to go through the lightbulb and then the transistor will be stopped by one of the PN junctions. In this diagram it is the upper one, which is a reverse-biased NP junction. At that time the vertical dashed line and the parenthesis are not in effect yet.

If a positive potential is now applied to the gate, it repels the holes in the P-type Si, causing it to become N-type (as indicated by the N in parenthesis). Now there is no PN junction directly in the path between the upper and lower N-type regions, because it is one long, continuous N-type region (drawn as a vertical bar, with the dashed-line as one side). This transistor is also N-channel, because the electricity goes through N-type silicon when it is turned on. Again N-channel is quite common, but P-channel devices are also used (in CMOS circuits, below).

The symbol for a MOSFET is shown on the right. The arrow in this case indicates that the source is internally connected to the substrate, which is done when the source PN junction is not going to be used. The substrate in this example forms a PN junction with the N-type drain, which can be reverse-biased.

If the device were P-channel, the source and drain would be P-type, and the arrow would be aimed away from the N-type substrate. This is shown at the top of the next diagram, Fig. 18.5.

CMOS

Two MOS transistors of opposite type can be wired as in Fig. 18.5, in the "complementary MOS" configuration ("CMOS"). *One of the transistors is always "off,"* so essentially-zero current goes straight down through the pair. Load current can be drawn from the output terminals, but in the situations when no such current is being used (the "resting" condition), *the overall current is practically zero.*

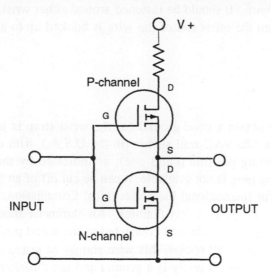

Fig. 18.5 Complementary MOS (CMOS) transistor pair.

Modern integrated circuits have millions of transistors attached in parallel, and even if only a microamp of "leakage current" flowed through each one that was not being used, an ampere or so would still be drawn from the power supply or battery at all times. This would generate a lot of heat and also drain batteries too fast for portable devices to be practical. Therefore almost all modern calculators, laptop computers, cellular phones, etc. use CMOS circuits whenever possible.

ESD Sensitivity

The MOS transistor is particularly susceptible to damage from high voltage "static electricity," of the kind generated by mechanical friction when a person walks across a rug in dry weather and makes a spark when later touching a grounded metal faceplate on a light switch. The spark is called "electrostatic discharge," or "ESD," but damage can be done even if there is no visible spark.

This voltage can cause dielectric breakdown (see index if necessary) of the very thin silicon oxide gate insulator, which destroys the device. Some MOS transistors are internally protected by Zener diodes that are attached in parallel with them, inside the packages, but most are not protected. To prevent damage, people handling IGFETs should always follow these two precautions: (1.) only touch plastic insulation, not the metal leadwires directly, and (2.) use a grounded wrist strap. The latter is a black or pink plastic strap that conducts electricity and is attached to a long wire. It should be fastened around either wrist, touching the person's skin, and then the other end of the wire is hooked up to a known good ground.

EXPERIMENT

The simplest way to obtain a good ground for the wrist strap is to put a three-conductor plug into a 120 VAC wall socket (in the U.S.A.), with only one wire screwed to its grounding pin (the round one), as indicated by the diagram on page 58. (If a 3-prong plug is not available, it can be cut off of an IEC extension cord. "IEC" stands for International Electrotechnical Commission, a cooperative industry group that specifies the safety features. An ohmmeter should be used to make sure that the green wire is indeed connected to the round pin.) Even when using a standard plug and wall socket, this wire should be tested with the neon bulb as on page 6, to be sure it really is a ground and not mistakenly wired to a "hot" connection inside the wall socket. The reason for such extra care is that wiring mistakes do happen (the author has found some), and a wrist strap is a good connection to the skin, so a person could be killed by such a combination of circumstances.

One or more clip leads can be strung in series to make the wrist strap wire long enough. Then the strap is fastened to the experimenter's arm. The circuit of Fig. 18.4 is then assembled, using an IGFET, quite analogous to the bipolar circuitry on page 169. Increasing the gate voltage causes the bulb to light, by enhancement of the substrate, which generates a continuous N-channel from the source all the way to the drain. The ammeter current is so close to zero that we can not measure it, even though the drain current is enough to light the bulb. Therefore the *current* amplification factor is enormous. The *voltage* amplification factor is good, but not so spectacular. (See also the curves in the Appendix.)

EQUIPMENT NOTES

Components For This Chapter	Radio Shack Catalog Number
Battery, nine volts	23-653
Clip leads, 14 inch, one set.	278-1156C
Lamp bulb, tungsten, 12 volt dc, green or blue	272-337A
Power MOSFET, N-channel	276-2072A
Static control wrist strap	276-2397A
Multimeter (or "multitester")	22-218 or 22-221
Neon test light, 90 volts	22-102 or 272-707
Power cord, IEC, with 3-prong plug	278-1258 or 278-1261

CHAPTER 19

Radio and Modulation

RADIO WAVES

An electric current going through a conductor generates a magnetic field, as already mentioned on pages 13 and 103 (and probably was known to the reader previously). A fast change in this field will generate a voltage in any stationary wire that is in the field, as mentioned on page 104. If the field is *very* quickly increasing and decreasing periodically, in other words it is a high frequency alternating field, there is something that might be surprising about the way this field goes a long distance as a "wave," and it can induce a voltage in a stationary wire that is far away. If the frequency is very roughly a million times per second (1 megahertz, or "MHz"), and it has been focused to form a parallel-sided beam, a moderately strong current in the original wire (the transmitting antenna) can send the beam 12,000 miles around the equator to be detected electrically in the other wire (the receiving antenna) on the other side of the world. This, of course, is a radio wave. The thing that makes the field strong enough to be detectable so far away is the extremely fast change in transmitting current, in order to go from zero to full intensity and back again in less than a millionth of a second. The effect would be much less at lower frequency, which is why radio usually involves MHz frequencies.

Although Heinrich Hertz is given credit for developing the early theory of radio and publishing his ideas, and Guglielmo Marconi operated the first commercially successful radio system (as mentioned on page 9), once again it seems as though Thomas A. Edison was among the first inventors. He obtained U.S. Patent 465,971 in 1891 which included pictures of transmission towers and even receiving antennas on ship masts. Edison had reported earlier experiments of transmission and reception in the magazine Scientific American, December 1875, and he predicted that radio would eventually replace telegraph wires. Hertz wrote about radio waves as being a new thing in 1886, without mentioning Edison. However, as happened with several other important ideas, Edison lost interest in radio, and instead he devoted his attention to inventing electric lights, phonographs, storage batteries, and the intermittent-motion movie projector.

EXPERIMENTS

Crystal Radio Receiver

The experimenter should build the circuit shown in Fig. 19.1. The diode can be a silicon 1N914, or whatever rectifier is available. A germanium diode such as a 1N34 will give better results because of the low forward voltage drop (page 153), but a silicon device will also work. The headphones should have a dc resistance of about 1K or more, because that indicates a lot of turns in its electromagnet coil, although lower resistances might also work. The types meant for pocket tape players usually are satisfactory, although the resistance should be measured with an ohmmeter, and higher is better, up to about 10K.

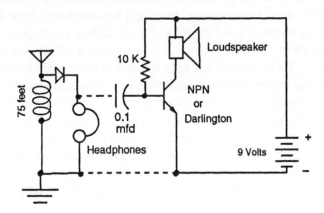

Fig. 19.1 Ultra-simple "crystal radio," with optional amplifier.

The antenna can be a 120 V extension cord, or preferably two, one being plugged into the other one in series, with a clip lead attached to one male plug prong. The ground should be the same as the type used for the wrist strap described in the previous chapter, with the same precautions being followed to make sure there is no power line voltage accidentally leaking to the "ground."

No capacitor is needed for the resonant tank circuit, because the "distributed capacitance" of the coil and other wires is sufficient for resonating at about 1 MHz (see index if necessary).

The coil of the resonant tank circuit, as usual in this course, has been simplified beyond the limits of what many people would believe might work, and yet it does work rather well. It is a reel of "hook-up wire," with an inside diameter of one inch and about 2 inches outside diameter. The coil is about 1/2 inch in "length" as defined on page 114. It is wound on a reel made of plastic. The length of the wire if completely uncoiled is 75 feet.

The ends of the wire have to be exposed, with the insulation stripped off enough for clip leads to grip bare metal, as described on page 65. If one end is buried inside the other coil turns, the whole coil might have to be unrolled to expose it and then coiled up again. At any rate, one end is attached to the ground and headphones with alligator clips, and the other end is attached to the antenna and diode. With the headphone against one or both ears, the wire is slowly uncoiled at one end only, with about 6 or 7 turns dangling toward the floor, still all attached to the alligators. Those loose turns are separated from the main coil by about two inches for each one, which decreases the inductance. As enough coils become separated, one or more local AM radio stations should become audible in the headphone.

If these additional parts are available, a loudspeaker, a battery and either a single transistor amplifier or a Darlington pair can be used to make the reception louder, by attaching wires where the dashed lines are shown in the diagram (still Fig. 19.1). The capacitor in the amplifier's input line is needed because it prevents the battery's *dc* from leaking back to the left and strongly reverse-biasing the diode. This is often done at the inputs of *ac* amplifiers. After attaching it, the coil might have to be readjusted to get back to the proper tuning.

This radio receiver is a "crystal set," with a "diode detector." The functions of each part will be described later in this chapter. The crystal in this case is the silicon inside the diode. However, as many readers already know, historically it used to be a single crystal of the mineral galena (lead sulfide), which is a natural semiconductor. It had to be contacted with a sharp metal wire. Surprisingly, the sharp point forces a small region of the N-type galena to become P-type, making a special kind of "point contact" PN junction. At any rate, old style or new, the detector is a rectifier. (It could be a three-wire transistor, with a battery, etc.)

Another optional variation, which might be tried after the silicon diode has been made to work properly by adjusting the coil inductance that feeds it with an RF signal, is to *make* a diode to replace the silicon 1N914. This can be done by using the tip of a steel needle or the corner of a razor blade as a "point contact." It is gently touched, at about a 30 degree angle, to either a brownish tarnished piece of copper such as a penny, or a dark gray tarnished piece of silver such as a dime. The discolored material is usually semiconducting oxide or sulfide, and a point contact to it can be an acceptable PN diode.

Damped Wave Transmitter

In the first chapter, a very primitive radio transmitter was described, and it was mentioned that it was similar to the one on the famous ship Titanic. The transmitter shown in Fig. 19.2 is closer to the Titanic's radio, because it has a capacitor in parallel with the inductor, making a tuned resonant tank circuit (see index if necessary). This concentrates the transmissions at one particular frequency range. Also, the oscillating relay repeats those transmissions automatically. If an antenna were connected, as on page 181, the radio waves could be heard on an AM radio farther away.

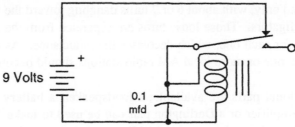

Fig. 19.2 Old-fashioned radio transmitter.

Although there is mainly one particular resonant frequency, there is really a fairly wide range of frequencies, because what is transmitted is a "damped wave," as illustrated in Fig. 19.3. When the relay contacts open during each cycle, the capacitor discharges through the coil, making an oscillation that is damped (see index if necessary) by the resistance. However, the system is not linear, so the frequency decreases with time, covering a range (note decreasing times between dashed lines). This makes it easier to receive the radio waves, because the system does not have to be exactly tuned to a single frequency. Of course, the Federal Communications Commission (FCC) frowns on such unlicensed and essentially random RFI. It is not recommended that an antenna be attached, because there might be illegal levels of RFI, as briefly discussed on page 183.

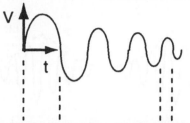

Fig. 19.3 Voltage versus time for a damped wave.

OPTIONAL

As an additional exercise, the circuit of Fig. 19.2 might be built and listened to with a portable AM radio. A louder signal will be heard, compared to the experiment in Chapter 1, because of the capacitor storing energy that then gets discharged within about one millisecond. Also, the crystal radio in Fig. 19.1 should be able to pick up the signals.

MODULATION AND DEMODULATION

Amplitude Modulated (AM) Transmission

The radio transmitter at the radio station involves a microphone to receive sound waves of very roughly 1 kHz frequency (see upper left corner of Table 19.1, on page 212) and convert these waves into an electrical analog of a sound wave. At the transmitting studio, when the sound intensity (really local compression of the air) is great, the electrical voltage is great, and when the sound intensity goes down to zero in the middle of a sine wave, the electrical voltage is zero. This should be a linear relationship, if possible.

Since electromagnetic radio waves need to have a much higher frequency (very roughly 1 MHz, as indicated in the upper middle part of Table 19.1), these waves are "modulated" by the electrical analog of a sound wave, and that is what is then transmitted by the radio station. Such a modulated waveform is shown in Fig. 19.4, with amplitude modulation (AM) at the top. The high frequency electromagnetic radio wave is very weak when the sound wave is weak, and has much stronger magnetism when the sound wave is strong. The radio signal is actually at a higher frequency compared to the sound wave than is shown in the diagram.

The radio wave is changed (modulated) according to the shape of the sound wave by changing its magnetic intensity (amplitude). Therefore this type of transmission is called "amplitude modulation."

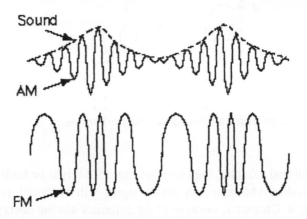

Fig. 19.4 Radio waves modulated by sound waves, AM and FM.

At a distant receiver, the magnetic radio wave in the air hits the antenna wire and is converted into an electrical signal at the same frequency. This goes into the coil, where the inductance plus the distributed capacitance have the right resonant frequency for the wave, and all other frequencies (that is, transmissions from other radio stations) are not resonant and are therefore very much weaker. This electrical signal then goes through the diode and the headphones, and then into the ground.

The radio frequency ("RF") is too high for the headphone or loudspeaker's moving parts to go back and forth with each of its up and down changes, so if unrectified RF was fed to them directly, they could hardly move at all. Therefore what is done in the crystal set is to rectify the RF, so only the top half of the waveform (as shown in the diagram) gets through the diode. Each time this pulsed dc increases quickly, it causes the headphone moving part to move a little bit, and the next time it gets moved a little bit more, but in the same direction. In that manner the headphone moving part does respond at the frequency of the sound wave, shown in the figure as a dashed line. (The dashed line is sometimes called the "envelope" of the waveform.) This generates a new sound wave in the air, which then travels to the listener's ear.

The radio transmitter has modulated the RF, and the radio receiver "demodulates" it, which is also called "detecting" it. Some radios can do either of these, and they can be called "transceivers." (They could also logically be called MODulator/DEModulatorS, or "modems." However, that term is reserved for circuits that convert computer outputs to telephone signals, and they can also convert those back into computer inputs at the other end of the telephone line.) Modern radio receivers have "detectors" that are transistors, instead of diodes, but they do similar things.

In case the reader is wondering about the bottom half of the wave, it can be represented by an out of phase RF wave, but otherwise the peak intensity of the radio signal is the same as for the upper half. With a simple diode receiver such as the one we will build, the bottom half of the wave is lost. However, the headphone's metal diaphragm tends to overswing and fill in the bottom half to some degree, and so does the human ear's vibrating skin membrane ("eardrum"). With a more complex receiver, there could be two transistors doing the job of extracting the audio frequency from the RF, instead of just a single diode, and one of these could operate on the bottom half of the signal.

AM signals are capable of high fidelity, but they are subject to noise interference, so radio stations generally don't bother to make their transmissions have the "hi-fi" freedom from distortion that FM stations try to achieve. Television signals use AM for the video portion and FM for the audio.

Frequency Modulated (FM) Transmission

As the reader probably has already figured out from Fig. 19.4, another way to modulate the RF magnetic wave is to change its frequency in a manner that is an analog of the original sound wave. The receiver must be more complex than with AM, and it will not be covered in this book. FM is very different from naturally generated magnetic waves in the atmosphere, so there is little background noise interfering with FM radio reception. On the other hand, AM is quite similar to many kinds of natural waves from lightning, sparks in electrical switches and other sources. Therefore AM reception is plagued with "static" noises.

Binary Digital Transmission

AM and FM are both linear analog technologies. If a bank wanted to send a telephone signal to a stock broker, saying that Joe Smith has $5 in his account, but the weather was hot and therefore the wires had a high electrical resistance, the signal might be slightly weakened, and it might communicate a current that is mistakenly analogous to only $4.99. It is quite difficult to send analog signals that are any more accurate than that, because of slight changes in resistance, and also from slight interferences due to natural background magnetic and electrical noises.

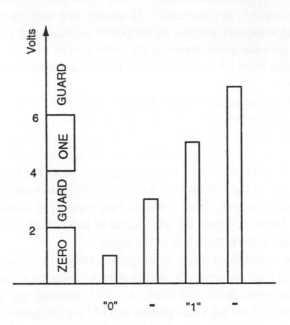

Fig. 19.5 Digital signal pulse heights and gate ranges.

By comparison, suppose that there could be a code that consists of electrical pulses, and the code for $5 is sent to the bank. This is what is done with digital technology. Looking at Fig. 19.5, an electrical pulse that is anywhere from zero to 2 volts high can be considered to mean "zero." If the pulse is something like 3 volts high, it is considered to be an error and is rejected. In fact, a pulse is then sent back to the transmitting system indicating that an error occurred and telling the system to try sending it again. If the pulse is from 4 to 6 volts high, it is considered to be a "one," but anything higher is an error. (These voltages are just examples, and real systems might have other "gate" values of voltage or current.)

The code can be binary, consisting of only zeroes and ones, as summarized by Fig. 19.6. In other words, $5 could be represented by two pulses separated by a space. If the pulses are only 4.99 volts high instead of 5.00 volts, it causes no error. In fact, they can be 4.1 volts or 5.9 volts, still without causing error. Also, it is unlikely that natural lightning, etc., would generate groups of pulses that looked like these.

Digital electronics can be billions of times more accurate than analog. Therefore, billions of operations can be done with essentially no error. Also, various operations can be done with digital coding such as "parity checks," which further reduce errors. The main **advantage** of digital technology is **accuracy** ("low noise").

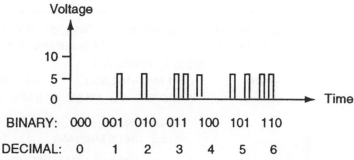

Fig. 19.6 An example of digital "pulse code modulation (PCM)."

A **disadvantage** of digital technology is that it requires equipment of great **complexity**. A digital radio receiver can not be made with just a loose coil and a diode. In fact, it requires hundreds of transistors.

However, modern integrated circuits (ICs) allow the use of millions of transistors (usually CMOS, as described in the previous chapter, so that very little current is needed), at low cost and taking up very little space. Thus **ICs** have overcome the problem of complexity, allowing us to utilize the accuracy of digital technology, and digital is slowly displacing analog (linear) technology.

At high frequency, more pulses can be transmitted per unit of time — that is, the information content is greater. Therefore the frequencies used in electronics are getting higher, as technologists learn how to transmit and receive them efficiently. Table 19.1 shows that satellite and cable TV, for example, are getting up into the gigahertz frequencies, and this seems to be the trend of the future: operations in the GHz range.

Because of the almost perfect accuracy, enormous amounts of digitally modulated information can be stored in small spaces. This is not because it is more efficient than analog methods — it is actually less efficient, in that it requires more operations to store the same amount of music, text, etc. However, the combination of photonic (Chapter 22) and integrated circuit (Chapter 23) technologies has allowed the inexpensive use of complexity crammed into small spaces, so using many operations to store or transmit data is no longer expensive. Some of the triumphs of these developments are apparent in Table 19.2.

Table 19.1   Frequency Spectrum

APPLICATION:	SOUND WAVES	AM RADIO (IF = 455 kHz)*	FM RADIO, TV (IF = 10 MHz)*	MICROWAVE

FREQUENCY:	Low Frequency 20 Hz to 20 kHz	Radio Freq.	Very High Freq. 54 to 216 MHz, FM 520 to 1,710 kHz, AM	Ultra-High F.
	~1 kHz	~1 MHz	~100 MHz	~1 GHz

EXAMPLES:	(Sound)	WNBC, 830 kHz	WNYC, 93.9 MHz	TV Satellite Downlink, 10 to 30 GHz
		Newark (NJ) Airport Tower, 118.3, 134.05 MHz (AM)		
			Channel 4 TV (VHF), 66-72 MHz	Ch. 40 TV (UHF), 626-632 MHz
			Cable TV, 5 MHz to 600 MHz	Cell Phones, 0.47 to 1.92 GHz
			Keyless Entry to Cars, 260 to 470 MHz	Bluetooth & Wi-Fi, 2.4 & 5 GHz
		International Short Wave (AM), 10.150 to 11.175 MHz		

RF Induction Heating, ~2.5 kHz (For conductors)	**Dielectric Heating, 27.12 MHz (For insulators)**	**Microwave Oven, 2.45 GHz (For wet objects)**

Global Positioning System, 1.2276 & 15.7542 GHz

Radio Astronomy, 37 and 74 MHz   Piscataway (NJ) Police Cars, 929.287 MHz (AM)

Police Radar, 10.50 to 10.55 GHz (X-Band)
24.05 to 24.25 GHz (K-Band)
33.40 to 36.0 GHz (Ka-Band)

RFID, 13.56 and 915 MHz

VELOCITY:	1,100 ft./sec. = 770 miles/hr (Atoms moving)	3×10^{10} cm/sec (Electromagnetism)		

WAVE LENGTH = λ = Velocity / Frequency

METRIC	1.5 cm to 15 meters	300 meters	3 meters	30 cm
ENGLISH	1/2 in. to 50 feet	300 yards	3 yards	1 foot

---

* The "IF" is intermediate frequency, for superheterodyne tuning.

The type of digital coding used most often for telephones, and also for CD and DVD music recordings, is "pulse code modulation" (or "PCM"), as shown in Fig. 19.6. A *single* space, occupied by *either* a zero or a one, is called a "bit." It takes eight of these to identify each alphabet letter or decimal number, being surely distinguished from all the others. Therefore an eight-bit cluster of bits, called a "byte," is needed for each character. This is the basis of modern digital technology, although other coding schemes such as PWM (to be described on page 236) are also used.

Table 19.2 Some Typical Examples of Information Content

---

8 bits = 1 byte = one letter or decimal digit
1 megabyte (MB) = 400 page book (only simple text, without graphics)
1 gigabyte (GB) = 1,000 book library (again, without graphics)
0.7 GB = one CD for music, one hour, high fidelity
5 GB = DVD disc, single layer (or 17 GB for multilayer on both sides)

IMPEDANCE MATCHING AND REFLECTIONS

On the backs of most TV receivers, there are screw-type terminals marked "300 ohms," which can be attached to a flat pair of conductors called a "twin lead," leading up to an antenna. An alternative antenna attachment is a socket marked "75 ohms," for use with a coaxial cable, instead of the twin-lead.

Those numbers represent "characteristic impedances," symbolized in electronics literature by Z_0. However, *they are not* the kind of impedance described near the top of page 101. Z_0 is a number which is useful for predicting whether there will be *reflections* when an electromagnetic wave goes from one transmission medium to another, such as a television signal going from an outside antenna cable to the wires inside the TV set. If you mistakenly hook the 300 ohm antenna wires to the 75 ohm internal wiring, the signal will get partially reflected at that interface and possibly bounce up to the antenna and back, causing "ghost" TV images. However, if the hook-ups are correct (300 to 300, or 75 to 75), there are practically no reflections. Similar problems could be present whenever RF generators such as magnetrons are attached to RF equipment such as microwave heaters. Therefore, some devices have knobs for adjusting the Z_0.

Even a straight wire has some slight inductance (briefly mentioned on page 103), and any *pair* of wires has some slight "distributed capacitance," (examples on pages 184 and 205). For a *simple* "transmission line" pair of wires, if L and C are known, they can be used to calculate the characteristic impedance: $Z_0 = (L/C)^{1/2}$. If an ac voltage is applied to an *infinitely long* transmission line, current will flow into it because of its capacitance, but inductance will limit the current so that $I = V/Z_0$. Most real systems require far more complex equations. In any event, the characteristic impedances should be determined by one means or another, and they should be "matching" (equal) whenever two "transmission lines" meet. Special instruments are used to detect undesirable reflections, by measuring the heights of the "standing waves" caused by those reflections, while various things get adjusted, to minimize the "voltage standing wave ratio (VSWR)."

Simple TV antennas are often designed to have Z_0 values of 300 ohms. For connection to a 75 ohm coax cable without reflections, there should be a special transformer called a "balun." The antenna is usually a "balanced line," while the coax is unbalanced, so this *isolation transformer* mates them ("balun" means "balanced-unbalanced"), but it also mates the *characteristic impedances*.

A mechanical analog of matching impedances is a rope, glued to the end of another rope, making one long cable that hangs down. Suddenly wiggling the top will send a wave to the bottom, *reflecting* back from the end, because the wave energy can not just disappear. However, if *either half* is replaced by a *thin string*, the wave will *reflect back from the glued joint.*

EQUIPMENT NOTES

Components For This Chapter	Radio Shack Catalog Number
Clip leads, 14 inch, one set.	278-1156C
Silicon diode	276-1103 or 276-1102
Wire, hook-up, 22 AWG, 75 ft.	278-1307 or 278-1221
Headphones	33-1100 or 33-1111
12 ft. extension cord	278-1261 or 278-1258
Capacitor, 0.1 mfd	272-1053 or similar
Relay, 12V dc coil	275-218C or 275-248
Portable AM/FM radio	12-794 or 12-799

OPTIONAL

Battery, nine volts	23-653
Resistor, 10K, (1/2 or 1/4 watt)	271-306 or 271-308
Darlington pair, NPN	276-2068
Loudspeaker	40-252 or 40-251

CHAPTER 20

Electric Motors

WHAT THEY ARE

DC Types

Because an automobile has a good supply of dc from the storage battery, there are many dc motors for such tasks as raising and lowering the windows, and many motor-like actuators for things like automatically regulating the air intake. Much larger dc motors are also used in industrial machines, elevators, and other activities involving a lot of starting and stopping, and changing of speed.

A simplified dc electric motor is diagrammed in Fig. 20.1. A coil "stator" is stationary, and a steel magnet "rotor" can rotate around the axle signified by a black dot. The half-circle is a copper metal "commutator," so the electromagnet

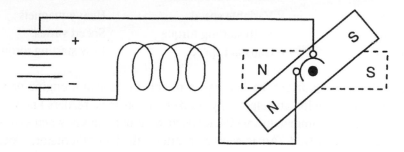

Fig. 20.1 A dc motor, with a mechanical commutator switch.

215

can be turned on in order to attract the "north" pole of the iron permanent magnet, which then swings up to the position shown by the dashed lines. In order to prevent this pole from tending to stop when it gets to that position, the conductive graphite "brushes" signified by the small white circles will lose contact with the commutator. The coil then loses its magnetism, and the rotor can continue spinning by inertia, until it comes around again to a position where it will be attracted when the commutator regains contact to the brushes.

The reader probably knows already that real motors have more coils and poles, and more complex commutators. There is more than one pulse of attraction, and sometimes the current is reversed to also cause repulsion. The coil almost always has a small soft iron core, and the permanent magnet is a large piece of hard steel or ferrite (iron oxide and barium oxide ceramic). Because the permanent magnet is heavier than the coil, it is usually the stationary part, not the way it is shown here. However, this diagram communicates the main ideas.

Other types of dc motors can have *coils* in both the rotor and stator magnets, which are able to provide more total magnetism than permanent magnets alone. The stator coil can be wired in series with the rotor coil, or in parallel (called "shunt" wiring). The main features of each type are listed in the following table.

Table 20.1 DC Motor Types

	Advantages	Disadvantages
Permanent Magnets	Variable speed	Heavy magnets
Series Coils	High starting torque	Speed varies
Shunt Coils	Constant speed	Low starting torque

An electric *generator* can be just like a motor, but a steam engine or water turbine causes the rotation and electricity is taken out, as most readers know. A dc generator requires a commutator, which tends to wear out quickly when extremely high currents are involved. An ac generator, often called an "alternator," does not need a commutator, so that type is used in modern automobiles.

AC Types

Almost all modern electric power for homes and factories is ac, mostly because its voltage can easily be "stepped up" by transformers, sent long distances with very little power loss, and then "stepped down" to 120V or 208V, etc., nearer to the user. The reason for such small power loss is that the heat loss is mainly

{Power lost in the wires:} $$P = I^2R \qquad (2.10)$$

as explained on page 17. Therefore decreasing the current will decrease the power a great deal. Because of the equation

{Power to local transformer and motor:} $$P = VI, \qquad (2.8)$$

increasing the voltage (to about 400,000 volts in long distance power lines nowadays) allows the current to be decreased by more than a factor of 1,000. Thus the heat loss during transmission is decreased by more than a factor of a million, compared to transmission at a few hundred volts (eqn. 2.10). Although the ac can be rectified to dc by the user, most electric motors operate on ac, because rectification always involves some equipment and also some power loss.

Repulsion Motors

It should be noted that, while an ac electromagnet will attract a piece of previously unmagnetised iron by inducing the opposite magnetic pole, a piece of nonmagnetic copper or aluminum will be repelled. Electricity has to first be induced in copper, and this generates its own magnetic field, which repels the original field. (Each turn of the extremely powerful magnets used in nuclear reactor experiments tends to repel each other turn, and such magnets would explode if they were not constructed with mechanically strong materials.)

If an ac motor has coils in both stator and rotor, it can be made with a commutator and will operate very much like the dc example described above. This is called a "repulsion motor," because attraction is not the main force used to get it started rotating, although both attraction and repulsion are usually involved in operation after it has come up to full speed. Sometimes these motors have extra windings, and some even use rectified dc to aid in starting or in speed variation. Some factory motors of such complexity are still being used. However, most modern electric motors are no longer designed with "hybrid" ac plus dc coils or with commutators for repulsion, because simpler designs have been improved to the point where they are competitive.

Induction Motors

The use of ac has an advantage in that no commutator is needed. The ac magnetic field of the coil (almost always with an iron core) will induce an ac current in any nearby conductor, so a copper or aluminum bar can be the rotor, as in Fig. 20.2.

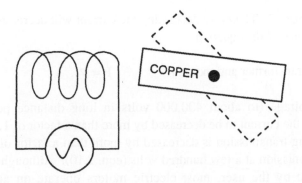

Fig. 20.2   An ac induction motor (synchronous), with a solid armature.

Very small motors such as this are often used in electric power meters, which the electricity supplier installs in a house or business to determine how much money to charge the user each month. This motor is not powerful, and it needs something else to get it started, which is not shown in the diagram.

Once it gets going, the rotor has to come around to the correct position for repulsion, just in time for the next ac cycle to occur. This is one of the principles of ac induction motors, and the speeds are usually "synchronous" with the 60 Hz ac. That is, the revolutions per minute (rpm) is 60 times per minute, divided by some factor that is dictated by the design, although there is a few percent of "slippage" behind an exactly synchronous speed. (Experiments with controlling the speed of an ac motor will be described in the next chapter.)

There are advantages to using an electrically conductive *coil* for the rotor, as shown in Fig. 20.3. This can provide much higher starting torque, because of the multiple turns in the winding.

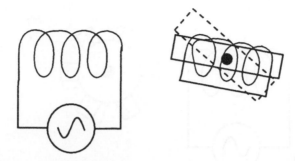

Fig. 20.3 An ac motor with a wire-wound armature.

Sometimes the incoming electricity is sent to the rotor coil, rather than the stator, in which case there is a metal ring in the middle of the rotor for making contact to the incoming wires, similar to a commutator. Such a ring is continuous, not divided into sectors like a commutator, and it is called a "slip ring" (not shown in these diagrams). Slip rings are also used in ac generators ("alternators").

Another way to have multiple conductors (but all wired in parallel as in Fig. 20.4) is called a "squirrel cage," which it really does resemble. Its advantages and disadvantages are also listed in Table 20.2 on page 225. It does not need any slip rings, because all the rotor current is induced.

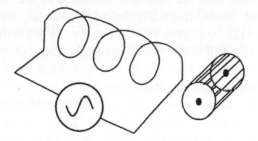

Fig. 20.4 An ac motor with a squirrel cage armature.

In order to have a synchronous motor run at slow speed, as in electric clocks, the rotor can have many poles, which is illustrated in Fig. 20.5. Clocks can have hundreds of tiny poles and turn at only a few rpm when driven by 60 Hz ac (3600 cycles per minute).

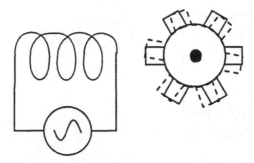

Fig. 20.5 A motor with a six-pole armature, 600 r.p.m. with 60 Hz ac.

Ordinary two-wire ac is called "single phase" electricity. Induction motors without commutators would have to be started mechanically (for example, by hand), if no special motor design were used with that kind of electric power source. Therefore some clever tricks are built into most motors to aid in starting.

Methods for Starting

An extra coil, usually quite small, can be arranged a few angular degrees from the main running coil. A fairly large capacitor is put in series with the coil, with its reactance being greater than the inductive reactance of the coil. Therefore the phase of this current "leads" (page 98) that of the strongly inductive main coil, which "lags" (page 112) the current in the other one. When starting, the rotor gets a little bit of temporary pull that is biased in one particular direction, and then the main coil keeps it going. Without that bias, it would just stand still, being pulled radially from the axle toward the stator, but not in any rotating direction. In a complex manner, the force vectors end up with the rotor *starting toward the capacitor and its small coil*. Once it has started, the rotor continues in that direction in a synchronous manner (Fig. 20.6). With a heavy mechanical load, the motor can continue to run, even if it falls far behind synchronous speed.

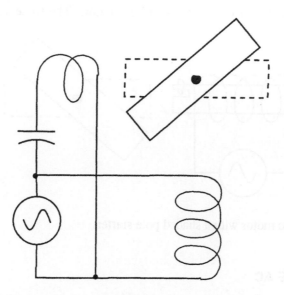

Fig. 20.6 An ac motor with a capacitor starter.

"Capacitor start" motors, listed in the table, run better without the extra little coil, once they are going. Therefore, some have centrifugal switches to disconnect the capacitor after they are up to a high speed. Small ones for such applications as cooling fans can keep the capacitor in the circuit continuously, since they do not use much total energy, so a small percentage that is wasted is not important. However, capacitors tend to become faulty sooner than coils, so these motors require more maintenance than some other designs.

A simple and reliable starting method, used by most modern fan motors and other low torque types is the "shaded pole" configuration. In Fig. 20.7, it can be seen that a small part of the metal or ferrite pole has a single turn of heavy copper wire around it. This acts like the short-circuited secondary coil of a transformer, as previously discussed on page 108. The inductance at that edge of the pole is decreased, similar to the effect of a small capacitor. The force vectors are quite complex.

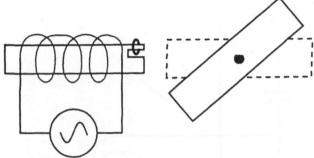

Fig. 20.7 An ac motor with a shaded pole starter.

THREE PHASE AC

The easiest and generally best way to start ac motors is to have the *magnetic field* of the *stator* be rotating. Then any conductive rotor design will be self-starting. This is one of the advantages of three-wire, three-phase electricity.

If the *generator* at the power station has *three coils* arranged as in Fig. 20.8, and there are three wires going from it to a *motor* that has *three stator coils* wired the same way, then the magnetic field in the motor will rotate at the same speed as the generator's rotor is turning. The voltage at each coil will reach a maximum value and then go down 60 times a second, just like ordinary two-wire ac (50 times a second in some other countries). The next coil will have the same ac in it, but one third of a cycle later, and the last coil will then get further-delayed ac. Much electric power is generated and transmitted this way, because of its great value for driving motors without needing capacitors, shorted turns, or commutators.

The original high voltage generator (480V or so), and the power company's "substation" transformer primary and secondary coils, can each be wired as in Fig. 20.8 (called "Y" or "Wye" or "star" wiring). Alternatively, *any* of them can be wired as in Fig. 20.9 on page 224 ("delta" wiring). A transformer *primary* can be wye and the *secondary* delta, or vice versa, or both can be the same type.

It is quite common for large electric motors and heaters to run on 208 volts. This can be obtained from substation transformer *208V* secondary windings of either type (wye, or delta), if the coil has the right number of turns. It can also be obtained from *240V* secondaries (but with delta, as explained later on page 224).

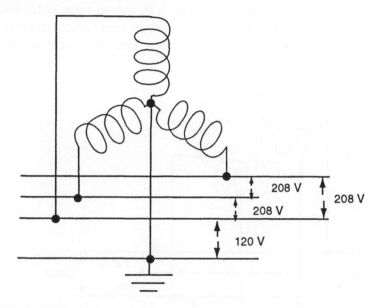

Fig. 20.8 Three phase transformer coils with Y ("wye" or "star") wiring.

When 120 volts is used for U.S. house or office wiring, it can be taken from the *middle* of the transformer's secondary windings, as shown in Fig. 20.8. Either a *fourth* wire can be used (for heavy currents), or else a very good *ground* is necessary, but neither are desirable for long distances.

Instead of that, 240 volts can be sent through the wires that go along the streets, either up on "telephone poles" or down in buried cables. Sometimes none of these are grounded. A *local* "step-down" transformer (not shown in the diagrams), usually only about a block away from the house or office building, has a center tap in its 240 V secondary that produces a *pair of 120 volt* circuits, with the center tap wire being shared. This center tap is the "neutral," and it is also *grounded locally*. For 120 volt service, sometimes one of the wire pairs is used for part of a building, with the other pair being used for another part of the same building, splitting the total load so that neither pair of wires has to pass too much current.

Another alternative for 120 volt service is to take it from a center-tapped delta type transformer, as shown in Fig. 20.9. However, where heavy usage of the 120 volts is likely, the other wires might then be unable to supply their full voltages, due to an "unbalancing" effect. Also, their phases (timing) might become altered and wrong for starting large motors. Electricity customers can discuss these alternatives with local power company engineers and find out what windings are available in the nearest substation. Sometimes a factory or large office building can buy its own substation and/or local center-tapped transformers and then only take high voltage inputs from the power company, usually at decreased cost per watt-hour.

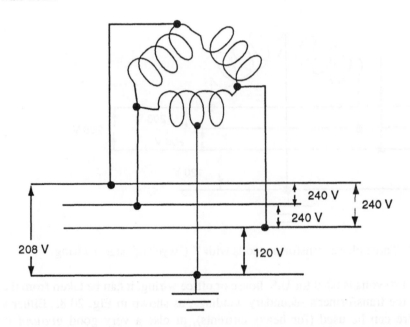

Fig. 20.9 A higher voltage transformer with center-tapped delta wiring.

Looking at Fig. 20.9, if the power company's generator is rotating clockwise (as shown in the figure), then let us consider the point in time when the upper-left coil's ac voltage is at its maximum value. At that particular time, the bottom (horizontal) coil voltage will be one third of a cycle "out of phase." Mathematically, it works out that because of this out-of-phase situation, the voltage between the *top of the triangle* (as shown here) and the *center tap* of the bottom coil gets 32 volts subtracted from the 240 volts, so 208 volts is available across those two terminals. (These are RMS values — see index if necessary.) This is one of the reasons why 208 volts is often used for electric motors and heaters.

Table 20.2 Comparisons of Various Configurations.

	Advantages	Disadvantages
Motor Winding		
Solid Armature	Inexpensive	Inefficient
Wire Wound	High starting torque	Expensive
Squirrel Cage	High running torque	Low starting torque
Multipole	Slow (for clocks, etc.)	Low torque
Starting Method, *Single* Phase		
Capacitor	High starting torque	Less reliable
Shaded Pole	Simple and reliable	Low starting torque
Repulsion	High starting torque	Needs commutator
Starting Method, *Three* Phase		
Wye	If close, 120V to ground	If far, 120V grounding
	Constant motor torque	
Delta	Three voltages available	Easily unbalanced
	Constant horsepower	

CHAPTER 21

SCRs and Triacs

THROTTLING AC CURRENT

It was pointed out that *high currents* can be controlled by reactance without losing energy as heat, back in the chapters on capacitors (page 102) and inductors (page 115). Another way to control high current is to *quickly* turn the current all the way on and then all the way off. This can be done as shown in Fig. 21.1, by using some kind of "gating" switch to stay turned off, and then suddenly turn on the current, but only when the voltage reaches a certain controllable percentage of its eventual peak value. More average current can be allowed to go through if the switch is turned on earlier, as in the right-hand diagram. Either way, at the right or left of the figure, no heat is lost ("dissipated") inside the switch itself, if it is working in an ideal all-the-way-on and all-the-way-off manner. By comparison, a rheostat would lose a great deal of power (maybe most of it) as heat.

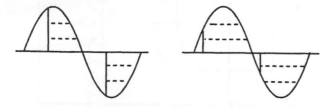

Fig. 21.1 Varying amounts of average current (dashed areas).

One type of semiconductor device, a "trigger diode," can be used to set the turn-on voltage, but it *can not handle the high currents* that would be useful for large heating elements or motors. Another device, an "SCR," can quickly gate the high currents, but it is *not easily adjusted* to turn on at certain desired voltage levels. Therefore both devices are attached together, and this combination is being used a lot for industrial electronic control systems, especially where high currents are required. (With small currents, it is still practical to use a rheostat as on page 47, or a transistor as on page 171, and simply waste the unused energy as heat.) More details of each device will be explained as we go along. First, however, we have to cover the basic ideas involved with turning on ("triggering") at some easily adjustable voltage, and then letting the electricity go further ("gating").

EXPERIMENTS

Complementary Monostable Pair

A good triggering device is a pair of transistors which is "complementary," that is, it combines PNP with NPN types. (The reader will probably note that this word was used before, with FETs, but there it meant the combination of N-channel and P-channel types.)

The circuit of Fig. 21.2 should be constructed. A fresh 9 volt battery, having about 9.3 volts when the voltmeter is connected to it directly, should provide

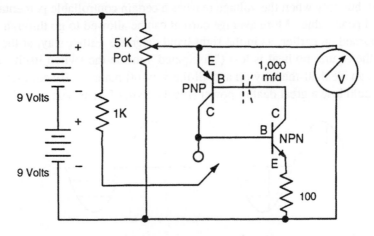

Fig. 21.2 Complementary monostable multivibrator ("trigger").

enough voltage for this experiment. A somewhat depleted battery with less voltage than that will require 2 batteries in series, as indicated in the figure.

With no capacitor, the pot is used to raise the voltage on the pair of transistors. At some point the voltage on the meter suddenly drops, indicating that both of the transistors have turned on. As the reader can figure out by inspecting the diagram, this is a positive feedback situation (a "virtuous circle," which is the opposite of a "vicious circle"), as we have seen before. The small amount of "leakage current" going through the initially "off" transistors finally reaches the point where it turns one of them "on" a little bit, and that turns on the other one even more, because of amplification.

Once the pair is turned on, lowering the pot voltage does not turn them off until it is lowered back down again, far below the turn-on voltage. They do go off again when the pot setting is lowered almost to zero, and then raising the pot voltage slowly again will show that they are still off, because the voltmeter goes to a high reading. This behavior, where the turn-on point is different from the turn-off point is called "hysteresis."

The voltage on turned-off transistors is often called V_{CC} (page 168). If it is set, via the pot, at a little bit below the turn-on point, and the 1K probe with its arrow is touched momentarily to the base connection of the NPN, that transistor will be turned on, and the whole pair will then go on and stay on, until the V_{CC} is lowered almost to zero. We will make more use of this idea in the discussion of the SCR.

NOTE: The following section should be studied, for general knowledge, even if the experiments are not performed.

OPTIONAL

If a capacitor is installed where it is shown, the pair of transistors will only stay on for a short time, regardless of whether the V_{CC} had been raised or the probe had been used. Therefore this is a monostable multivibrator, similar to the one on page 189, but this time using complementary transistors. This circuit is another one that can be useful for generating short pulses. It can also be used to convert a slowly rising voltage like a sine wave to a fast rising pulse like a square wave. One possible application of that is a "debouncer," in case switch contacts produce noisy on-and-off initial voltages (as on pages 4 and 135), and a small capacitor will convert those multiple spikes to a single, smooth square wave.

OPTIONAL
Schmitt Trigger.
Another circuit that is often used to remove contact bounce noise is the Schmitt
trigger, which is shown in Fig. 21.3. This has nicely adjustable hysteresis,
without requiring two kinds of transistors. (Note that electronics designers
often refer to transistors as Q1, Q2, etc.) Initially, Q2 is on and Q1 is off.
When Q1 is turned on by raising the pot setting (simulating an incoming

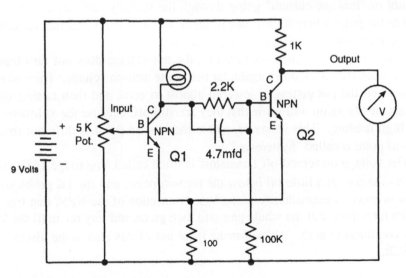

Fig. 21.3 Schmitt trigger hysteresis circuit.

sine wave or noisy contact voltage), it diverts current from Q2, turning it off.
The triggering voltage necessary to do that can be adjusted by changing the
values of the 1K and the 100 ohm resistors, since they make up a potentiometer, when Q2 is on. They determine the voltage that will be present at the
top of the 100 (which is called the emitter voltage, or V_E). In order to drive
current into the base of Q1, the pot has to supply a slightly higher voltage than
that V_E. That will be the setting that can switch things over to the other side,
far enough to make Q1 turn on.

Once Q1 is on, it tends to short-circuit the voltage on the base of Q2, so the
5K variable pot now has to be turned way down to the point where the
resistance of Q1 goes up, which can then drive current into the base of Q2. (If
a 56K resistor is substituted instead of the 100K, then Q2 would not go on until
an even lower setting of the variable pot is reached, giving more hysteresis.)

The capacitor does not limit the time that Q2 is on, because current can flow around it ("bypass" it) through the 2.2K. The purpose of the capacitor is, instead, to "speed up" the switching action.

MONOLITHIC DEVICES

PNPN Trigger Diode

The two transistors of Fig. 21.2 can be combined in a "monolithic" device made out of a single piece of silicon, which is sometimes called a trigger diode, sometimes a Shockley diode, and sometimes a PNPN. As illustrated at the left side of Fig. 21.4 (without the dashed lines), it acts as though two transistors were glued together side by side but displaced by one step. The symbol is at the upper middle of the figure. If a voltage above a certain amount is put across it, from the top-left P to the bottom-right N, it will "break down" to an "on" condition of low resistance, just like the two transistors did.

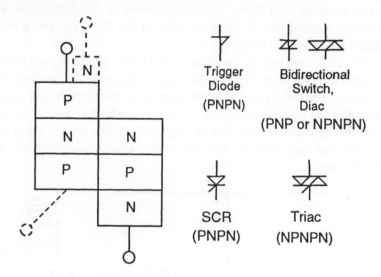

Trigger
Diode
(PNPN)

Bidirectional
Switch,
Diac
(PNP or NPNPN)

SCR
(PNPN)

Triac
(NPNPN)

Fig. 21.4 Various trigger and gate devices.

Silicon Control Rectifier (SCR)

If a third wire is attached to a PNPN, as indicated by the lower left diagonal dashed line, this triode acts like the two transistors of Fig. 21.2, with the new wire being like the terminal that was touched by the 1K probe. That is, if the voltage across the whole device is not quite enough to cause breakdown, and a small positive voltage is contacted to the third terminal, the device will then break down and stay at this new low resistance until the current is somehow brought almost to zero, after which it will turn off again.

This triode is called a silicon control rectifier, or SCR. (It is sometimes called a "thyristor," getting that name from the thyratron vacuum tube that it replaced.) It is the "gated switch" referred to in the first paragraph of this chapter. The symbol used in circuit diagrams is at the lower middle of the figure. The triggering device that can turn it on is usually a PNPN diode. With a potentiometer, the voltage that is used to turn on the PNPN can be adjusted, so the SCR itself turns on near the beginning of a sine wave, or else at nearly the peak.

Diacs and Triacs

SCRs are still being used in older equipment, but they are gradually being replaced by similar devices that work in both directions, + and –, to operate on both the top and the bottom of an ac sine wave. First of all, the PNPN trigger diode can be replaced by a "diac," which will work in either direction, + or –. Two alternative symbols are at the upper right of the Fig. 21.4, and it is sometimes called a bidirectional switch.

The SCR can be replaced by the "triac," which also works in both directions, and its symbol is at the lower right of the figure. A typical circuit diagram is in Fig. 21.5 below. The battery can be replaced by an ac transformer, and depending on the setting of the potentiometer, the waveform at the bulb would be just like those of the first illustration, Fig. 21.1 on page 227. When the voltage gets up to a certain point, the diac turns on, and that turns on the triac, which can handle very high currents.

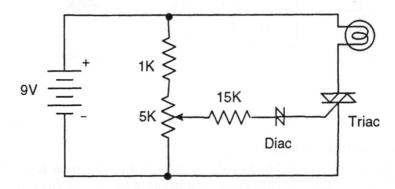

Fig. 21.5 Triggering and gating circuit.

The diac can be made with either three "layers," PNP, or with five, NPNPN. The former type is simply forced into avalanche like a Zener diode, and its characteristic curve was shown on page 153. (The PNPN diode's curve looks like half of that, with the other half being just a vertical line of high voltage but no current.) The triac requires an extra "layer" of N-type silicon that only covers part of the underlying P-type material, as shown at the top of Fig. 21.4. It operates something like an SCR or the two-transistor circuit that we made in the previous experimental section, on page 228, but it works in both directions.

EXPERIMENTS, PART 2

Relaxation Oscillators

The diac can be studied with the circuit of Fig. 21.6, but using the dashed-line wire and leaving out the transformer for a while. The voltmeter is attached across the diac. When the pot voltage reaches the diac's trigger level, about 6 volts, the meter will go to nearly zero. The capacitor will discharge through the 100 ohm resistor.

This type of circuit is used to make repeating sawtooth waves (see index if necessary) to drive the horizontal sweep of an oscilloscope or TV tube, but in the simplified version shown here, it is only good for a single pulse, because it stays on. The reason it does not go off is that the current is still fairly high, enough to

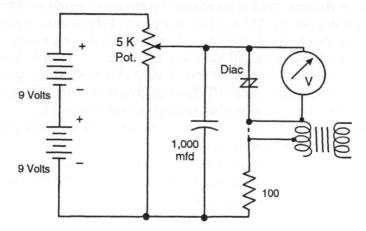

Fig. 21.6 Relaxation oscillator, with a diac, turned off by overswing.

keep it going, even after the capacitor has fully discharged, which is not supposed to happen. We have only a small voltage available, less than would ordinarily be used, so our resistor (the pot) has to be small, for the capacitor to charge up in a reasonable time. With that small resistance, current continues to flow, preventing the diac from turning off. (If there had been a higher voltage supply and thus a high resistance pot and a very small capacitor, then the diac would turn off.)

In order to get this diac to turn off, we will have to do something clever. (This is an example of how problems can be solved by simple means, with modern electronics.) An inductive coil (the 6 volt center tap of the 12 volt transformer) is put into the circuit instead of the dashed-line wire. The experimenter should do that now, and then set the pot just high enough to switch the diac, after a few seconds delay for charging up the capacitor.

Without further adjustment of the pot, the meter will periodically go up slowly and then go down fast, showing that sawtooth waves are being created. What the inductance does is overswing the current, causing the capacitor to charge a *very small amount* in the reverse direction (+ at the bottom), making a slightly oscillating tank circuit. This turns off the diac, since there is not enough + voltage to turn it back on in that direction.

OPTIONAL

Another way to get the diac to turn off is to put a slight "reverse bias" on its lower electrode. This is done in Fig. 21.7, on the next page. Instead of having the lower electrode "look at the ground" (as electronic designers often say), it is attached to a pot (the 330 and 100) that puts a slight positive bias on it. The capacitor slowly charges up through the 10K to the full voltage. Then it discharges, bringing the left-hand wire of the diac to "ground" level voltage (close to zero). The right-hand wire of the diac is now slightly more positive than that, so the diac turns off. The capacitor then charges up again, going through the cycle, and generating another sawtooth wave.

The top of the sawtooth is not straight, since the capacitor charges in a curved, exponential manner, as shown back on page 94. Therefore, for use in making horizontal sweeps for cathode ray tubes, a constant current diode is usually put in series with the resistor (10K as shown in this diagram), to force the charging curve to become linear.

Power Controllers

The main use of these devices is to control large amounts of power, for heaters, lights, and electric motors. The circuit of Fig. 21.8 should be assembled. The plug *should be attached to a ground fault interrupter* (see index if necessary).

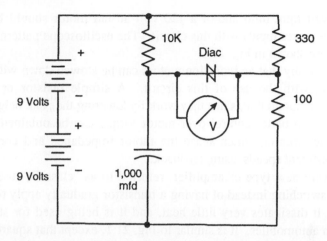

Fig. 21.7 Relaxation oscillator, turning off the diac with reverse bias.

Care must be taken not to touch any of the "hot" wires (upper wires in Fig. 21.8), once the power is plugged in. Also, care must be taken to ensure that the "ground" of the scope is really attached to the "neutral" of the 120 volt ac, in order to prevent tripping the ground fault interrupter and possibly damaging the scope. A neon tester can be used to check this, as on page 6. The potentiometer can be taped to the workbench so that it only needs to be touched with one hand, once the power cord has been plugged in. Although this experiment is dangerous, it will provide good experience for students, and it can all be done safely.

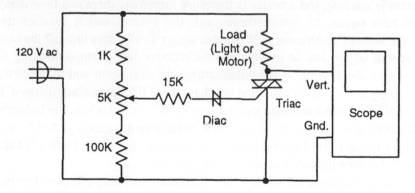

Fig. 21.8 Controlling power with a triac. (Bottom wire is "neutral.")

A 100 watt light bulb and/or a 120 volt ac fan motor should be throttled (decreased in power input) with this circuit. The oscilloscope patterns should be similar to those shown in Fig. 21.1.

It is noteworthy that ac induction motors can be slowed down without losing much torque, with the use of this circuit. A simple resistor or a variable transformer can not do that, since they work by lowering the peak height of the ac waveform. Even better control over motor torque can be obtained with more complex triac circuits, which sense the motor impedance and correct for its changes at different speeds, using feedback.

A relatively new type of amplifier, referred to as "class D," uses very fast on-off triac switching instead of having a transistor gradually apply resistance to the output. It dissipates very little heat, and it is being used for stereo music amplifiers in automobiles. It is similar to Fig. 21.1, except that square waves are used (having all the same heights, but various widths). This is called "pulse width modulation (PWM)," or "one-bit coding." A relatively new type of voltage regulating power supply, referred to as a "switcher," similarly "chops" the output into pulses of various widths or various repetition times, depending on the power needed. It is very efficient, though expensive, and it is used for aerospace, etc.

PROPORTIONAL CONTROLLERS

Home heating systems, air conditioners, and refrigerators ordinarily have their temperature regulating done by "on-off" (sometimes called by engineers "bang-bang") switches of some kind. For example, in Fig. 21.9 on the next page, if a temperature is too low, and a heater is therefore turned on, there is a time delay until the heat begins to spread throughout the system and it reaches the temperature sensor ("thermostat"). When this sensor finally does turn off the heat source, it will be too late to prevent some excessive heat from continuing to spread through the system, so the temperature will overswing and temporarily become too high. This is shown in the top diagram of the figure ("not damped"). It is an oscillation, like electronic oscillations in LC circuits where the inductor causes overswing of voltage in the capacitor, even after the transistor has turned off the input (page 179). In general, it is partly caused by a time lag (or "phase difference") between the power source and the sensor.

There are two convenient ways to decrease the overswing effect in heating systems. The one used in most *home* systems is called an "anticipator." A very small heater is placed right next to the sensor, so when the heat is turned on, the sensor responds sooner than it would otherwise, thus decreasing the time lag. It has also been used in laboratory and factory equipment.* While this makes the problem less bad, it does not completely eliminate overswing.

A better way to attack the problem, used in most modern *engineering* temperature controllers, is called "proportional control." In this system, the amount of heat (or cooling effect in a refrigerator) is decreased as the temperature gets closer and closer to the desired value. In other words, the power applied is proportional to the "error signal" that the sensor is indicating. This proportionality is the "P" in modern systems which are referred to as "PID controllers."

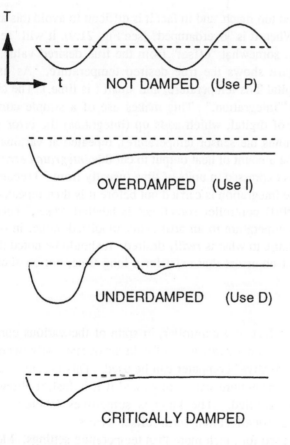

Fig. 21.9 The effects of adding P, I, and D actions to a controller.

---

* Daniel Shanefield, "Improved Thermostat System," *Review of Scientific Instruments*, Vol. 32, 1961, pages 211 and 1403.

The Effect of P

The proportionality (which usually requires fairly complex electronics) has the overall effect of "damping" the controller system, and this decreases the overswing. Sometimes the knob on a controller that adds this function is labeled "damping," but most often it is called "proportional band (PB)." If the "band" is made narrower, there is a steeper gradient of power increase as the temperature goes down; if the band is set by the operator to be wider, there is a more gradual application of power (but over a wider temperature range).

The Effect of I

The damping can be used too much, and in fact it is difficult to avoid this and still stop the oscillations. When it is "overdamped" (see Fig. 21.9), it will "settle" at the wrong temperature, somewhat "offset" from the true desired value. (The dashed line in the figure shows the true desired temperature. As usual in scientific diagrams, capital T is temperature, and small t is time.) The cure for this problem is to use "integration." This makes use of a simple computer, usually analog instead of digital, which adds up (integrates) the error signals (desired temperature minus the sensor temperature), repeated at various times, and it *slowly* changes the amount of heat output to cut this *integrated error* down to zero. Sometimes it is expressed in units of time (usually from 30 seconds to 2 minutes) over which the integration is carried out before it is then repeated. The integration knob on a PID controller sometimes is labeled "reset," because it changes the "setting" temperature to an artificially modified value, in order to slowly drift the temperature to what is *really* desired. It should be noted that this knob only has an effect on errors that exist for a long time, not on short-term "upsets."

The Effect of D

Sometimes the overall effect of a controller, in spite of the various correction factors, still overswings when a *short* blast of cold air occasionally occurs, etc. To minimize this, a "derivative" computer can be used. This is sensitive to the *slope* of the measured temperature versus time, and if it is fast, it allows more corrective effect to be applied. The knob is sometimes labeled "D" and sometimes "rate." It does not affect long-term "offset" errors.

PID controllers are used for much more than temperature settings. More and more automation machinery is controlled by such equipment, which prevents the motion of robot arms from oscillating, and minimizes cumulative error, etc. Some factory workers fail to understand the principles outlined above and simply "twiddle the knobs" almost randomly, hoping to get good control by luck.

NOTE:

Usually the best sequence of setting these controllers is to adjust P first, then I, and then D, taking considerable time to let things reach a constant value before making the next change.

EQUIPMENT NOTES

Components For This Chapter	Radio Shack Catalog Number
Battery, nine volts *(possibly two)*	23-653 or 23-553
Clip leads, 14 inch, one set.	278-1156C or 278-1157
Lamp bulb, tungsten, 12 volt dc, green or blue	272-337A
Multimeter (or "multitester")	22-218
Transformer, 120V to 12V, 450 ma, center tapped	273-1365A or 273-1352
Resistor, 100 ohm, 10 or 1 watt	271-135 or 271-152A
Resistors, 330, 1K, 2.2K, 15K, 100K	271-306 or 271-308
Potentiometer, 5K	271-1714 or 271-1720
Capacitor, electrolytic, 1000 mfd	272-1019 or 272-1032
Transistor, PNP	276-1604 or 276-2043
Transistor, NPN *(two required)*	276-1617 or 276-2041
Diac ("bidirectional switch"), 6 volts	900-3156 or 900 3157 *
Triac, 400V	276-1000
Cooling fan, 3 inch, 120VAC	273-242 or similar
Neon test light, 90 volts	22-102 or 272-707
AC power cord, 6 ft.	278-1255 or 278-1253
Tape, electrical	64-2348 or 64-2349
Enclosure box	270-1809
Ground fault interrupter	(from hardware store)

Oscilloscope, Model OS-5020, LG Precision Co., Ltd., of Seoul, Korea

OPTIONAL

Capacitor, electrolytic, 4.7 mfd	271-998
Resistors, 10K, 56K (or 47K), 100K	271-306 or 271-308
100 watt light bulb and socket	(from hardware store)

---

* These and similar items can be purchased from the Radio Shack mail order subsidiary, RadioShack.com, P. O. Box 1981, Fort Worth, TX 76101-1981.

CHAPTER 22

Photonics

WHAT IT IS

From quantum mechanics (see page 142 if necessary), we have learned that the rules of nature only allow whole number multiples (such as 1x, 2x, etc.) of the smallest units of energy, not fractional multiples (such as 1.3x, 2.78x, etc.). A single unit of energy of electromagnetic waves such as light is called a "photon." Quantum mechanics and relativity, two powerful systems of knowledge, seem to describe separate realms of nature for the most part, only overlapping in a few places, and sometimes apparently disagreeing (or at least competing). Albert Einstein, discoverer of the rules of relativity, did not have any love for quantum mechanics, and in fact he had philosophical disagreements with some aspects of it. However, although it is not well known, Einstein actually got the Nobel Prize for discovering the "quantized photoelectric effect," which was the keystone that supported the early theories of quantum mechanics. (Surprisingly, he did *not* get the prize for relativity, because the judges thought his theory might be incorrect!)

The smallest unit of *sound* wave energy (the motions of atoms — see bottom of page 212) is the *phonon*. In some ways the phonon, and also the electron and *photon*, can each behave like a wave, and in some ways each can behave like a little particle.

Just as electrons can travel long distances through a wire, photons can travel under the whole Atlantic Ocean through a glass fiber. An important modern trend is to communicate via optical fibers, and therefore we need to be able to convert electrical energy into light, and back again. This chapter will cover some of those conversions.

EXPERIMENTS

Photons to Electrons and Back Again

On pages 150 and 160 it was mentioned that some PN junction diodes can give off light when a forward current flows, as the electrons "fall into" the holes. However, on page 146 it was briefly mentioned that the opposite can also happen: light can give its energy to a valence band electron, and that can then have enough energy to go up into an empty space in the conduction band of a semiconductor, which will become electrically conductive. The experimenter should *attach a glass-packaged 1N914 diode to the oscilloscope and expose the diode to bright light.* Using the most sensitive scope setting, a measurable dc voltage appears.

Next, we can make even more electricity with a very efficient "solar cell," which is a large PN junction near the surface of a silicon plate. The *red* wire of a silicon solar cell is attached, via clip leads, to the *longest* wire of the LED (light emitting diode), which is the leadwire of the *P-type* region of a gallium arsenide-phosphide PN junction. The solar cell is exposed to very strong light, preferably direct sunlight, or else a 100 watt lightbulb up close. The LED will give off visible red light, if that lamp is placed inside a dark paper tube to shield it from the sunlight. What is happening is shown in Fig. 22.1.

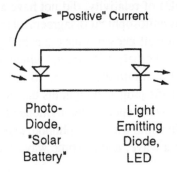

Fig. 22.1 Light to electricity to light.

Some solar cells are so efficient that they can drive incandescent lamps, heating the tungsten filaments hot enough to emit visible light. Attaching the 25 ma lamp in place of the LED can demonstrate that. Of course, a lot of energy is lost as heat in each conversion step, since they are not 100% efficient, so the light source has to be very strong.

PN junctions are not necessary for these processes. At lower efficiency, light can raise the energy level of an electron, pushing that up to a conduction band, and thus allowing electrical conduction. This usually does *not generate* electricity like a good quality PN junction can in a solar cell, but it does allow a *separate battery* to push electricity through an illuminated material that would be an insulator without the light shining on it. The phenomenon is called "photoconduction." In a TV studio's transmitting camera, light from an actor hits a screen and knocks off electrons, which are collected and then sent to an amplifier, and that process was briefly described on page 87.

In a different mode, individual electrons flying through a vacuum can hit a material and give their energy to the material's electrons, which then fall back into holes and give off light. In the cathode ray tube (CRT) of an oscilloscope or a TV set, the electrons coming from the hot cathode and hitting the screen, then emitting light, provide an effect called "cathodoluminescence."

Electrons to Photons and Back Again

Although we will not build this sort of device, it should be close to obvious that one circuit could be used to drive a light emitting diode, and the light from it could drive any kind of light-sensitive diode which is in a completely separate circuit. This "optically coupled" pair of diodes is called an "opto-isolator" (sometimes spelled without the hyphen). It can be used to achieve "isolation" (see page 82 if necessary) so there is no common ground wire, thus eliminating the chance of a "ground loop" (see page 61 if necessary). It works better than a transformer when very high frequencies are involved, because it has no inductance. Also, for safety purposes, it can be used to isolate a high voltage circuit from the rest of an electronic system.

Two other uses for the optically coupled pair of diodes involve having an opaque object occasionally interrupt the light beam. One common application (using invisible infrared wavelengths) is at the entrance to stores, to alert the sales people to customers coming in or out of the building. This can also be used to operate elevator doors or to trigger burglar alarms.

Another application is in a "chopper," where the opaque object is similar to the blades of an electric fan or the blades of a ship's propeller. A motor rotates the blade assembly, and each blade interrupts the light beam for a short time, thus generating electric "square waves" in the second half of the system. This arrangement is used in a "lock-in amplifier," to help minimize background noise effects, especially when low power light beams are being studied, as is often done in fiber optics research. During the time that the light beam is *hitting* the opaque blade, the only electricity in the light-sensitive diode half of the system is background noise, from room light or RFI or other random sources. This can be *subtracted* from the electricity that flows when the blade is *not* in the way of the beam, by using a "differential amplifier" (to be described in the next chapter, on page 250) or by using a digital computer (as in the Chapter 24). Either way, analog or digital, the signal that remains after the subtraction does not include that noise. An excellent level of noise reduction is usually achieved this way (in other words, an increase in the "signal-to-noise ratio.")

Photon-To-Photon Conversions

Similar quantum processes can convert light into other light, having somewhat lower energy. A familiar example is fluorescence. High energy ultraviolet light can be absorbed by a material that has the right electron energy levels available. The photon disappears, giving its energy to an electron, which then falls into a hole and re-emits light of lower energy (usually visible light).

The emission does not have to take place immediately. Complex rules of quantum mechanics describe situations where the electron that has been kicked up to a higher level might stay there. Quite often this involves impurities of the same general kind that "dope" semiconductors — that is, they are in an energy gap. The light being absorbed might produce either holes or trapped electrons (stuck in acceptor atoms that are not near others). The dopants can be purposely diffused into materials to make this effect more likely, and that is done commercially to make "phosphors." If light strikes the phosphor and "pumps" electrons (or holes) into higher energy positions, it might take a long time for new light to be emitted, and that would be called "phosphorescence," in contrast to immediate emission, which is "fluorescence." This is the phenomenon used to make light-switch face-plates, etc., glow in the dark.

When a lot of electrons have been pumped up into unnaturally high energy levels, this is called "population inversion." They usually wait until some other energy comes along at the same level, and then for some strange reason in nature, that "stimulates" the original energy to join the newly-arrived energy, and both come out together. Sometimes the newly-arrived energy is just the random heat from several moving atoms, in which case the phosphorescence slowly emits a glow of low energy light.

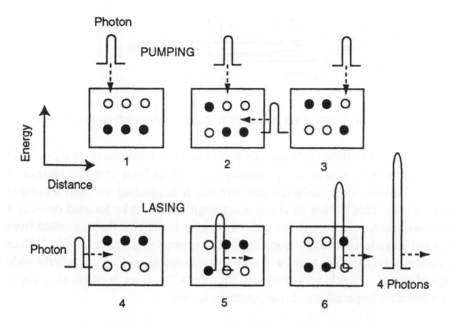

Fig. 22.2 Light amplification by stimulated emission (a laser).

On the other hand, if the newly-arrived energy is in the form of a light wave, it can hit an electron that is waiting at a unnaturally high level, and then the two energies will become joined as a doubly-high energy pulse of emitted light. This "amplifies" the original light (doubling its intensity in this case), and it is described as "light amplified by stimulated emission," or LASE. A laser is a device that does exactly that. In Fig. 22.2, the process is shown in simplified diagram form. Pumping from the side causes population inversion (picture number 4), followed by newly-arrived light at the end, which gets amplified before it comes out. Since a crystal of sapphire or ruby (aluminum oxide with various dopants) has trillions of pumpable atoms, the amplification factor can be extremely high.

To multiply things even further, mirrors can be arranged as in Fig. 22.3, so that some of the light goes back and forth many times before finally getting out through the half-silvered mirror at one end.

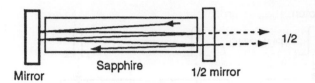

Fig. 22.3   A laser, intensified by multiple passes through the material.

Pumping can be done electronically by PN junctions, instead of by light. That is what happens in a laser diode, usually made of gallium arsenide instead of silicon. Various other materials are also used, depending on their available energy levels. This light is all at one wavelength, so it can be focused down to a very small point, compared to the wide range of wavelengths that is emitted from and a hot incandescent lamp. Because of "dispersion" (various refractive indices for various wavelengths), a lens will only focus incandescent light to a fairly wide range of points. That is why lasers are used for CDs, laser printers, etc., where high resolution (separability of fine points) is needed.

Photocopiers and Laser Printers

The principal of the dry photocopier is illustrated in Fig. 22.4. A cylinder is covered with photoconductive material such as arsenic selenide. In the dark, a high voltage is used to spray electrons onto the cylinder, and they stay there in the form of a static electricity charge, because the material is an insulator in the dark.

Light reflected off a piece of newspaper (not shown) is focused onto the cylinder. Wherever the paper is dark, because of ink marks, no light is reflected, but where it is clean, there is corresponding light. This is represented in the diagram by a rectangular barrier in one area, but that would not actually be present.

Where the light hits the photoconductor, the static charge gets conducted to the ground and is no longer present. Then black powder particles, sometimes wax particles containing carbon black, are sprinkled onto the cylinder. They only stick where there is static electricity, similar to dust sticking to a phonograph record whenever it is statically charged on a dry day.

The cylinder gets rotated, and its charge is then reversed. The black particles are repelled and fall onto a clean sheet of paper. They are then melted by a heater and become permanent black marks on the paper.

In some variations of the process, positively charged air molecules are used instead of electrons, and sometimes magnetic iron oxide (black) is used for a magnetized cylinder process, and magnetism is even combined with static electricity in other variations. This process was developed during a period of twenty frustrating years (!) by Chester F. Carlson, but it finally became a huge technological and financial success.

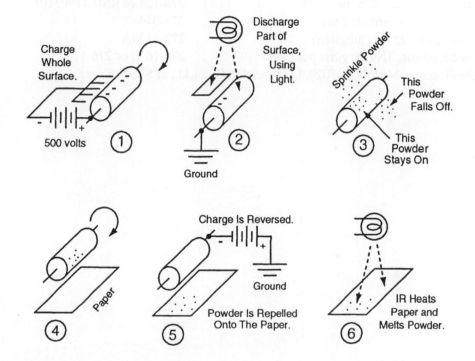

Fig. 22.4 The main ideas behind photocopiers and laser printers.

A laser printer for computers works almost the same way. A laser sends finely focused light to the cylinder, turned on or off for each tiny spot ("pixel"). The black powder ("toner") can be the same material, and the paper can also be the same. Together, the photocopier and laser printer have made a veritable revolution in the efficiency of office work.

EQUIPMENT NOTES

Components For This Chapter	Radio Shack Catalog Number
Clip leads, 14 inch, one set.	278-1156C
Silicon solar cell, 275 ma	276-124 or RSU 11903101
LED, red, low current, 2 ma	276-044
Mini-Lamp, 25 ma (tungsten)	272-1139A
Diode, silicon, 1N914, *glass package*	276-1620 or 276-1122

Oscilloscope, Model OS-5020, LG Precision Co., Ltd., of Seoul, Korea,

CHAPTER 23

Analog Op-Amp ICs

INTEGRATED CIRCUITS

An integrated circuit ("IC") consists of many "components" (such as resistors, capacitors, transistors, etc.), where *many of the parts* of these components have been made simultaneously, *in a single operation.* In other words, if a microprocessor IC for a small computer contains 2 million transistors, all 2 million of the emitters are made at the same time, in a single step, and then all 2 million bases are made in one step, and so on. Once they are made (by diffusion, etc.), they become part of a single ("integrated") piece of material — they are not separate things soldered together.

ICs have three important advantages over "discrete component" assemblies. They are very much

more reliable, and
cheaper, and
smaller.

The first advantage is the most important, because computers with 2 million transistors would be very unlikely to work at all unless each device had a very low probability of failure. However, the other two advantage are obviously important also, when millions of components are involved. This technology has a great effect on our daily lives, along with some other topics discussed in this book.

OPERATIONAL AMPLIFIERS

A large group of transistors can be combined in a single circuit, to provide a *huge amplification factor*. Also, this can be *adjusted* over a wide range. That makes it easy to apply either positive or negative *feedback*, by having a wire go from the output back to the adjusting input. Such an amplifier circuit is called an "operational amplifier," or "op-amp." It can be made from discrete components, but nowadays it is usually available as an IC.

Differential Amplifiers

One of the first inexpensive op-amp ICs was the LM741, and many newer ones are variations of that classic design. The input is a pair of transistors which make up a "differential amplifier," shown in Fig. 23.1. "Positive current" going into the *"negative"* input *lowers* the output voltage, by nearly short circuiting it to ground (if the emitter resistor is small). However, the *"positive"* input allows more current to go through the common emitter resistor, making more voltage drop across it, and therefore *raising* the output voltage (assuming that Q2 is *partially* turned on, which is usually the case, because of a bias resistor that is not shown here).

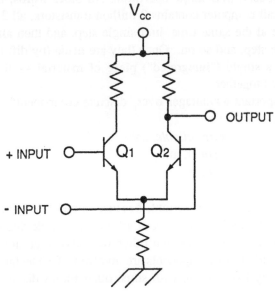

Fig. 23.1 A differential amplifier.

This differential amp input stage makes it easy to apply feedback, as briefly mentioned above, because the "feedback loop" can go back as shown in Fig. 23.2. The triangle is the symbol for an amplifier.* A resistor adjusts the amount of feedback, with a small one providing more effect, and a very large resistance providing only a very slight feedback effect. (Negative feedback is often used in high fidelity audio amplifiers, in order to decrease distortion, making the *analog-type* output-versus-input relationship be *linear*.)

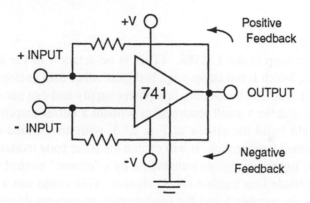

Fig. 23.2 A commonly used op-amp, with two kinds of feedback.

Another advantage of having a differential amp as the input stage is that a two-wire, "balanced" input (page 80) becomes inherently able to reject RFI and other EMI noise (see index if necessary). Such noise will induce the same voltage on both input wires, which will cancel itself. This is quite useful when op-amps are magnifying very weak signals, by means of their really huge amplification factors (a million or so). Otherwise, the noise would be greatly amplified, along with the signal, making all that amplification worthless. The ability to eliminate this kind of noise is called "common mode rejection ratio (CMRR)."

The 741 needs two power supplies, one with a + voltage relative to the ground (or the chassis), and one with a – voltage, as shown in the figure. Although this is a disadvantage, it is being made available more often in modern power supplies, because op-amps are so useful. The 741 is an IC containing 20 bipolar transistors. It is not capable of delivering much output current and usually needs a buffer.

---

* Sometimes the left-hand vertical line of the triangle is curved.

Op-amps such as the 741 and similar designs are rapidly replacing individual transistors in circuit diagrams that the reader is likely to see and wish to use. Newer types are available with FET transistors, and some contain both FET high-resistance inputs and high-current bipolar outputs. Generally, op-amps are the "building blocks" of many industrial circuits, and they should become another part of the reader's "intellectual toolbox."

EXPERIMENTS

The 386 Op-Amp

An easier-to-use op-amp is the LM386. (This is no relation to the Intel 386 micro-processor IC, which is not an op-amp, is digital rather than analog, and has CMOS transistors.) The 386 needs only one power supply and can put out more current — enough to drive a small loudspeaker without a buffer amplifier. The experimenter should build the circuit of Fig. 23.3, with an infrared-sensitive photodiode and a small loudspeaker. It will enable the pulse code modulated (see index if necessary) lightwave signals transmitted by a "remote" control for a TV or DVD set to be made into audible sound signals. (We could use a 4.7 mfd capacitor between pin number 5 and the loudspeaker, to prevent dc output, but that is not necessary in this simplified circuit.)

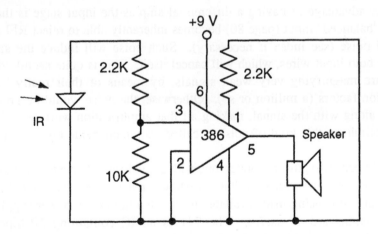

Fig. 23.3 The LM386 op-amp, making TV remote signals audible.

The 386 is in the form of a "dual in-line package (DIP)." The black material is injection-molded epoxy polymer, with carbon black added to prevent random light from putting stray voltages on the PN junctions of the IC. Two lines of leadwires stick out of the bottom, like the legs on a centipede insect. It is difficult to attach clip leads to these, so a "breadboard" socket module will be used. This will provide useful experience for the experimenter, since these modules can be helpful in future circuit building.

Looking at the top of the DIP, with the leadwires facing away from the experimenter, the circuit diagram of what is inside the package is shown in the next diagram, Fig. 23.4. The pin numbers identifying the leadwires should be compared to those numbers in Fig. 23.3.

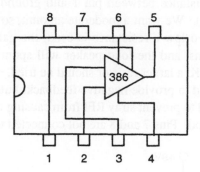

Fig. 23.4 Looking down onto the top of a 386 in a plastic DIP.

Breadboard Module

Before the days of ICs, experimenters often wired discrete components by mounting them temporarily on a wooden board of the type used for slicing a loaf of bread. (Those "good old days" are clearly remembered by the author, who built his first crystal set — and thought of improvement modifications — in 1938. At that time, bread was usually purchased in the form of a whole loaf and then sliced at home. The author borrowed his mother's wooden breadboard for radio experiments and was punished for getting it dirty and full of nail holes.)

The plastic breadboard module we will use has wires buried inside. Small spring-loaded sockets are also inside, wherever the square holes are visible. In Fig. 23.5, a long row of horizontal holes is all connected together, to one long "bus" wire. The bottom one can be used for "ground" connections (really just going to the negative battery terminal). The top long row can be used for the battery positive connection.

A short row of five holes (four vertical dashes in the figure shown here) is all connected to a buried vertical wire. Each other row is similarly connected to its own wire.

A jumper wire of the right length can be taken from a kit of different sizes, or it can be made by cutting a piece of thin solid 22-gauge copper wire. Leadwires on resistors are bent to a U-shape, so each end can be poked into the breadboard holes. Thus the input wire and the 2.2K are connected together via a vertical row of holes. The other end of that 2.2K, and the 10K, and a wire going to pin 3 are all connected together via another row of holes. Input and output wires can be attached to clip leads.

The lower the resistance between pin 1 and ground, the higher the amplification factor ("gain"). We want a moderate amount, so 2.2K is appropriate for this application. If too low a resistance is used, the gain is so great that stray noise starts oscillations, and the loudspeaker will spontaneously howl. If this happens even with 2.2K, a larger resistor should be tried.

Pin 2 could be used to provide negative feedback, but we will not use it here, so it must be grounded to prevent stray RFI from causing trouble. Pin 4 should be the equivalent of ground. Pins 7 and 8 are not connected to anything.

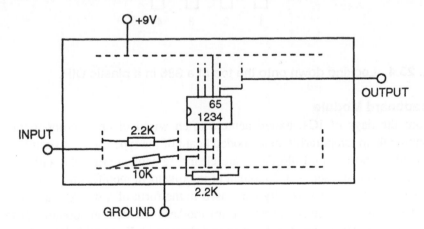

Fig. 23.5 Looking down onto a breadboard, used for wiring Fig. 23.3.

If a Radio Shack catalog number 276-142 photodiode is used, the package looks like the one on page167 of this book. What is shown as the emitter on that page becomes the small triangle of the diode on page 252, where "positive current" comes out when light goes into the PN junction. The middle wire of the photodiode package is attached to the small horizontal line of the diode on page 252, where electrons come out. The infrared ("IR") light that is detected goes sideways into the transparent package's curved-surface lens, not into the end. (It would be instructive to look at the chip and "wire bond" connections inside the package, using a magnifying glass.)

A remote controller for a TV or DVD is aimed at the photodiode, up close, and operated. Pushing various function buttons should cause various clicking sounds to be heard via the loudspeaker. This is digital pulse code modulation, as in Fig. 19.6 on page 211, similar to the coding used in CD or DVD music recordings.

OPTIONAL

The IR light from the remote controller can be aimed at an infrared sensor card, in a dark room, or in a darkened space under an overcoat. The IR is made visible by a special kind of fluorescence called an "anti-Stokes transition."

EQUIPMENT NOTES

Components For This Chapter	Radio Shack Catalog Number
Battery, nine volts	23-653
Clip leads, 14 inch, one set.	278-1156C
LM386N Operational Amplifier IC	276-1731
Infrared detector	276-142 or 276-145
Experimenter socket ("breadboard")	276-175 or 174
Jumper wire Kit	176-173 or RSU11642238*
Resistors, 2.2K (1/2 or 1/4 watt) *(two each)*	271-306 or 271-308
Resistors, 10K (1/2 or 1/4 watt)	271-306 or 271-308
Loudspeaker	40-252 or 40-251

OPTIONAL	
Infrared sensor	276-1099
Magnifying glass	63-848 or 63-851

---

* See asterisk footnote on page 262.

CHAPTER 24

Digital Microprocessor ICs

WHAT THEY ARE

Because binary digital coding is so error-resistant (page 211), it is used for the various internal operations of most modern computers. Analog electrical signals are still used for many purposes such as local telephone calls, and for some types of fairly simple computers in automation, etc. However, relatively small but very powerful desktop computers, and also some large supercomputers, now use digital "microprocessors" for their internal operations. In each of these, most of the action takes place on one silicon IC "chip." A large single crystal of silicon is grown by slow cooling from a melted pool, and this is later sliced to form a "wafer." That gets donor or acceptor atoms diffused into various regions of it, making mostly CMOS transistor pairs. Then the wafer is diamond-sawed to make hundreds of smaller "chips," each one of which is almost a whole computer. Of course, it also needs to be attached to a keyboard, and a TV-type screen, etc.

The internal operation of such devices is beyond the scope of this quite simplified laboratory course. Suffice it to say that the binary digital programs for these microprocessors consist of extremely long lists of binary (zero or one) numbers. This is called machine language.

Groups of these numbers, somewhat summarized into a symbolic format, are called object code, and this is considered to be a sort of "language," at a slightly "higher level" than the raw zeros and ones. Programs for running the computer can be written in object code, as long as there is another "object linking and embedding (OLE)" program also in the computer's memory.

257

At another slightly "higher level" is "assembly language," in which programs can be written using far fewer symbols. These, however, must be "compiled" into raw machine language before the computer can use them, and that takes a little bit of extra time and is less efficient regarding memory.

At a still higher level of summarization, programs can easily be written by people who have only limited training in computer operation, if they use languages such as BASIC or FORTRAN. The highest level languages are the ones involving a "graphic interface," where a mouse is used to click on "icons," or a person simply typewrites with a word processor program. These are the least efficient for the computer, but for most people they are the easiest ones to use.

Whatever language is used to write a program, that language is referred to as the "source code." For professional programmers it is often assembly language.

EXPERIMENTS

Reading Low Level Data Inside the IC Chips

The experimenter should have access to an IBM-compatible laptop or desktop computer. With the computer turned on and "booted up," the DOS prompt is obtained on the screen (possibly by going into the "Programs" directory and selecting "PC DOS Prompt"). The experimenter then types:

DEBUG

and presses the RETURN or ENTER key. A short dash symbol shows on the screen, which is the DEBUG program's prompt.

Then the experimenter types

r

on the keyboard and again presses RETURN. This command tells the computer to display the contents of the "register," which is a temporary memory section in the "central processing unit," or "CPU." That is the main IC in the computer, which might be the 80486 type, or a more modern chip. The screen will show a series of things such as

AX=0000 BX=1980 ...etc...   NV UP NA   ...etc...
DS=1568 ES=FFEE ...etc...   PUSH BX   MOV SP, [SI] ...etc...

The AX and EX and similar two-letter symbols represent *locations* in the register, and after the equal sign comes the *data* that is temporarily stored there, depending on what particular operations the computer had been told to do just before it was told to run DEBUG. These data are not binary, but instead they are in the "hexadecimal" numbering system that the computer uses in its fairly low level (that is, fairly close to the ultimate binary) operations. Instead of using only the binary zeroes and ones or decimal zeroes to nines, hexadecimal uses zero up through nine, and then continues with alphabetical letters, A through F, counting up to the equivalent of decimal 15. After that, it has to go up to the next sequence, equivalent to *16* ("hexadecimal" in Latin). This is just like a decimal number, after running out of digits after number nine, having to add another digit (the zero), in order to go up to the next sequence, decimal *10*. (Thus, decimal 9 = hex. 9, dec. *10* = hex. *A*, dec.15 = hex F, dec. *16* = hex. *10*, dec. 17 = hex. 11.)

The short words, like PUSH and MOV, are "assembly language" for various operations which the computer had to perform previously. Assembly language uses *alphabetical letter* abbreviations, but it uses *hexadecimal numbers*.

The next thing to do is type

d

and press RETURN. An array of numbers appears on the screen. The command means "dump" in a low level language similar to assembly language, and it shows the contents of the first part of the random access memory ("RAM"), which is usually not part of the CPU but is in a separate chip ("expandable" memory). The first four digits displayed, plus the next four after the colon, will show the location of a whole line of flip-flop transistors in the chip. The conditions of these flip-flops are displayed in hexadecimal, because that is the language of the computer's operating system, as made visible with DEBUG. Typing q will quit the DEBUG program and go back to the DOS prompt.

Measuring a Resistance with BASIC

With the computer *turned off* completely, a joystick is plugged into a socket that fits its connector. Inside the joystick base are two rheostats (variable resistors) of about 100K each. Moving the joystick handle changes one or both of the resistor's settings, depending on the axis of motion and the angle. The computer sends a voltage out through the socket and measures the time that is taken to charge up a capacitor inside the computer, going through the joystick's resistor. The longer the time, the higher the resistance that is sensed by this action. The amount of time taken to reach about 5 volts is then stored in one of the computer's RAM memory transistors.

With the computer turned on and "booted up," a file searching program[†] is used to find QBASIC or ABASIC or BASIC, which is probably in the C:\WINDOWS directory or in the original CD-ROM. Opening that file, or else typing those letters and then hitting the RETURN key, BASIC will run. If a "help" screen shows first, press the ESC key to bypass it. With the cursor in the upper, "Untitled" window, the experimenter should then type

    10 PRINT STICK(1)   (then press RETURN)
    20 GOTO 10

and press the RETURN key again. The program is started by hitting ALT and R simultaneously to get the RUN menu, and then hitting S for "start." (In some computers, hit the SHIFT and F5 keys simultaneously.) Moving the joystick handle positions will show different numbers on the computer monitor.

The BASIC program can be stopped in various ways, depending on the age of the computer. Old models will stop with FUNC and DEL or ALT and ESC. Newer ones might need CTRL and BREAK to stop the operation. Using (2) instead of (1) in the BASIC program will measure the angle of the joystick handle along a different axis.

To exit BASIC, use ALT and F and then X. To exit the DOS prompt, type EXIT and hit RETURN.

If the joystick base can be taken apart easily, an external resistor can be substituted for the 100K rheostat (carefully, to avoid short circuits!), and experiments can be done to correlate various resistances with the screen displays.

This is one type of analog-to-digital ("A-to-D") conversion, changing an analog resistance value to a digital number, and there are many different types. In this case a number increases, step by step at every beat of a clock, and when a limiting voltage is reached, the number at that point (such as "86") is shown on the screen. Quite often the A-to-D conversion (or a D-to-A conversion) is the least accurate part of the process, and that probably is the case here. But it is accurate enough for games, and it could be done in a better manner if necessary.

EQUIPMENT NOTES

Components For This Chapter	Radio Shack Catalog Number
Joystick, Avenger 700	26-377
IBM-compatible computer	

[†] For example, with the DOS prompt visible, type: dir/s/p ?basic.* to find QBASIC or ABASIC.

Equipment List for Entire Course

	Radio Shack * Catalog Number
Battery, nine volts †	23-653 or 23-553
Battery, AA, 1.5 volts	23-872 or 23-882
Clip leads, 14 inch, one set.	278-1156 or 278-1157
Neon test light, 90 volts	22-102 or 272-707
Transformer, 120V to 12V, 450 ma, center tapped	273-1365 or 273-1352
Portable AM/FM radio	12-794 or 12-799
Multimeter (or "multitester") †	22-218 or 22-221
Lamp bulb, tungsten, 12 volt dc, green or blue	272-337 or B or C
Fuse, fast-action, 315 ma, 5 x 20mm †	270-1046 or 270-1047
Phillips head screwdriver, for replacing fuse	64-1950 or 64-1901
Color code slide rule	271-1210
Resistor, 100 ohm, 10 or 1 watt	71-135 or 271-152A
Resistors, 150, 180, (1/2 or 1/4 watt)	271-306 or 271-308
Resistors, 330, 1K, 2.2K (1/2 or 1/4 W) *(two each)*	271-306 or 271-308
Resistors, 15K, 56K, & 100K, 1/2 or 1/4 watt	271-306 or 271-308
Resistors, 10K (1/2 or 1/4 watt)	271-306 or 271-308
Potentiometer, 5K ††	271-1714 or 271-1720
AC power cord, 6 ft.	278-1255 or 278-1253
Power cord, IEC, with 3-prong plug	278-1258 or 278-1261
12 ft. extension cord	278-1261 or 278-1258
Wire, hook-up, 22 AWG, 75 ft.	278-1307 or 278-1221
Wire stripper	64-2129A
Soldering gun, 30 watt	64-2066A
Solder, flux core	64-018
Tape, electrical	64-2348 or 64-2349
Enclosure box	270-1809
Headphones	33-1100 or 33-1111
Relay, 12V dc coil	275-218C or 275-248

Oscilloscope, Model OS-5020, LG Precision Co., Ltd., of Seoul, Korea, with offices also in Cerritos, CA, USA.

---

*,† Footnotes are on next page.

Diode, silicon, 1N914, glass package*(two each)*	276-1620 or 276-1122
Diode, Zener, 6V, 1N4735, 1W	276-561 or 276-565
Diac ("bidirectional switch"), 6 volts	900-3156 or 900 3157 *
Triac, 400V	276-1000
Cooling fan, 3 inch, 120VAC	273-242 or similar
Piezo transducer	273-073A or 273-091
Loudspeaker	40-252 or 40-251
Transistor, PNP	276-1604 or 276-2043
Transistor, NPN *(two each)* †	276-1617 or 276-2041
Darlington pair, NPN	276-2068
Power MOSFET, N-channel transistor	276-2072A
Static control wrist strap	276-2397A
Capacitor, electrolytic, 1000 mfd *(two each)*	272-1019 or 272-1032
Capacitor, electrolytic, 4.7 mfd *(two each)*	271-998
Capacitor, 0.1 mfd *(two each)*	272-1053 or similar
Capacitor, 0.01 mfd *(two each)*	272-1065 or 272-1051
Infrared detector diode	276-142 or 277-1201
Infrared sensor	276-1099
Experimenter socket (breadboard)	276-175 or 174 or 169
Jumper wire kit	176-173 or rsu11642238*
Low current red LED	276-044
Mini-Lamp, 25 ma (tungsten)	272-1139A
Silicon solar cell, 275 ma	276-124 or RSU 11903101
Magnifying glass	63-848 or 63-851
LM386N operational amplifier IC	276-1731
Joystick, Avenger 700	26-377
IBM-compatible computer	

---

* These and similar items can be purchased from the Radio Shack mail order subsidiary, RadioShack.com, P. O. Box 1981, Fort Worth, TX 76101-1981, phone 1(800) 442-7221 ,
e-mail commsales@radioshack.com,
website www.radioshack.com.

† Several extra units should be made available, because they are easily damaged.

†† NOTE: It would be a good idea to draw a short line on the metal shaft of the 5K pot with a marking pen, so its rotated position is always apparent.

Although it is not necessary, it would be a good idea for the instructor of this course to purchase a "ground fault interrupter" (see page 57 and index) in a hardware store, attach a power cord (similar to the item above), and place it and its socket in a plastic "enclosure" box such as Radio Shack Catalog Number 270-1809. Then all 120 volt ac power, even for the soldering iron, can be obtained via this safety device.

ADDITIONAL READING

Much more can be learned about many of the topics in this book by looking them up in the book listed on page 10.

APPENDIX

Silver	1.6×10^{-8} ohm-meters
	(1.6×10^{-2} ohm-centimeters)
Copper (99.99%)	1.7
(More impurities are usually in copper, raising the resistivity.)	
Aluminum	2.7
Nickel	6.1
Chromium	13.2
80%Ni, 20%Cr, Nichrome	102.2
(In Index, see also "nickel-chromium alloy" and "scattering.")	
Silicon (99.99999%)	3 million
	(Extremely sensitive to impurities.)

Measured by making a cube of the material, one meter on each side, and putting ery conductive metal such as silver on two opposite faces, to serve as electrodes. An ohmmeter is attached to those electrodes, and the resistance will then also be the esistivity. If the material was longer, but still the same width and thickness, then he resistance would be greater, but the resistivity would still be the same, since it is nly defined for a cube. Similarly, if the material was wider or thicker, the esistance would be less. This is somewhat analogous to density, which is defined or a cubic centimeter, and a larger volume of that material would weigh more but till have the same density as a small volume.

Of course, it is just a coincidence that the measurement unit, *ohm-meter*, looks imilar to the word for the instrument, *ohmmeter*. It should be noted that older ables often list the resistivities in ohm-*centi*meters, rather than in ohm-meters.

DIELECTRIC CONSTANTS

As described on page 92, this is the k' or "relative dielectric constant," measured at 1 kilohertz. Since it is relative to the k of vacuum, it has no dimensions.

Vacuum	1
Polyethylene plastic	2.3
Water	78.5 (Sensitive to impurities)
Ethyl alcohol (ethanol)	24.3
Aluminum oxide ceramic	8 to 10
Silica glass	3.8
Mica mineral	2.5 to 8.5
Titanium oxide	~100
Barium titanate ceramic	~1500
Lead magnesium niobate ceramic	~20,000 (Sensitive to oxygen content)

DIELECTRIC STRENGTHS

As described on page 146, this is the resistance to avalanche conduction ("dielectric breakdown") in an insulator. A mil is a thousandth of an inch, or about 25 micrometers ("microns"). All the values below are approximate, since they are highly dependent on humidity, defects, rate of voltage increase, and other factors.

Glass, typical	~400 volts per mil of thickness
Silica glass	~600
Porcelain ceramic	200 to 400
Aluminum oxide ceramc	~9,000
Barium titanate ceramic	~10,000
Mica mineral	1,000 to 10,000
Polyethylene plastic	~500
Polystyrene plastic	~700

MAGNETIC PERMEABILITIES

As described on page 114, this is the μ or "relative permeability," measured at 1 kilohertz, at low magnetic field. Since it is relative to the μ of vacuum, it has no dimensions.

Vacuum   1

Iron   150 to >5,000
(Very dependent on impurities, annealing, etc.)

Transformer Steel, 97% Fe, 3% Si   1,400
(Silicon raises the electrical resistance, thus lowering the losses due to "eddy currents" induced in the steel itself, which would have decreased the induced power available in the secondary wire windings. Also, the steel is used in thin sheets, with insulating paper in between, to further interfere with eddy currents. That is why the symbol used in drawings is three parallel lines, as shown on page 3. However, the eddy currents are still too lossy for use at radio frequencies, above about 40 kHz.)

Permalloy, 45% Ni + 55% Fe   2,600

$Ni_{0.3}Zn_{0.7}Fe_2O_4$ ferrite ceramic   ~8,000

$Mn_{0.6}Zn_{0.2}Fe_{2.2}O_4$ ferrite ceramic   ~2000
(Can be made electrically *insulating,* for low loss at radio frequencies. Small amounts of silicon dioxide, etc., are put in the grain boundaries, and also the composition is further optimized, to raise the electrical resistance and thus lower the losses due to RF-induced eddy currents.)

VACUUM TUBE CHARACTERISTICS, Typical Diagram

The following curves show one of the ways that vacuum tube characteristics are presented in the manufacturer's specification sheets, for a typical radio tube, a 12A5 pentode. (The 12 is the filament voltage, and the 5 shows that, in order to produce high current, it has 2 more electrodes than just the 3 electrodes shown on page 193.) High voltage is needed for the plate. In England, vacuum tubes are called "valves."

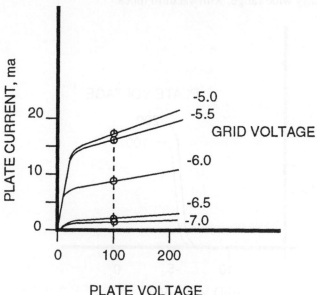

These curves are shown quite often, because they allow the experienced designer to use a diagonal "load line" for finding the currents that would flow with various output resistances. (An example is explained on page 271.) However, this type of plot is not very intuitive for inexperienced users, who would rather have a characteristic curve such as the one on page 173, showing *input versus output* directly. In order to obtain that format, the vertical dashed line shown here at 100 volts can be drawn by the user, and the data at each small circle can be read off the graphs. These can then be plotted as shown on the next page, for each plate voltage, which does give a family of *input* (grid voltage) *versus output* (plate current) curves.

VACUUM TUBE CHARACTERISTICS, Intuitive Diagram

Here the "saturation" at high currents can be seen more easily, at the tops of the curves. The characteristics are most-linear (least-distorted) in the middle region of each curve. The useful amplification factor is the slope of that linear portion, which is the change in output current with change in input (grid) voltage. Ordinarily, the plate voltage is not changed during use, but the circuit designer has the option of setting it over a fairly wide range, with vacuum tubes.

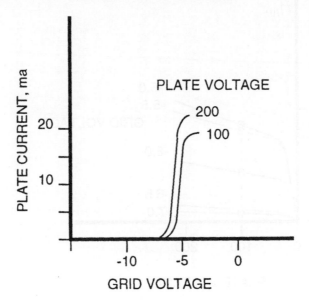

FIELD EFFECT TRANSISTOR (FET) CHARACTERISTICS

Typical FET "spec sheets" use formats similar to those of vacuum tubes. Useful amplification is obtainable with much lower driving voltage (V_{DS}). Analogous to the tube situation, the *input* is the V_{GS} and the *output* is the I_D. In this case a type 2N7000 MOSFET is used in N-channel enhancement mode, with a common source circuit as on page 198.

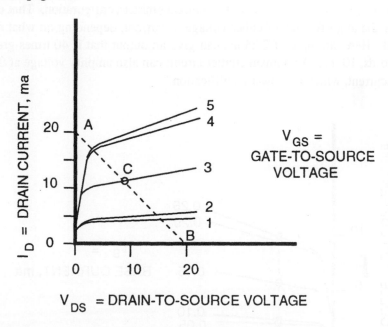

$V_{GS} =$
GATE-TO-SOURCE
VOLTAGE

V_{DS} = DRAIN-TO-SOURCE VOLTAGE

A "load line," as briefly mentioned on page 269, is shown here as a dashed line. Its slope represents the effect of a certain load resistance (in the same place as the lightbulb is shown on page 198). In the case graphed above, that resistance is 1,000 ohms, and the V_{DS} is 20 volts. The dashed load line is drawn from the maximum voltage point (shown here as B) to the maximum possible current point with that particular load, which is $20V/1K\Omega = 20ma$ (shown as A). If the transistor is turned partly "on" ($V_{GS} = 3$ volts), the drain current would be about 11 ma, as shown by the intersection (the circle under the letter C).

BIPOLAR TRANSISTOR CHARACTERISTICS

Typical bipolar transistor data are similarly presented, as follows. In this case a type 2N525 transistor (NPN) is used in the common emitter configuration. That could be done for the amplification of either voltage or current, depending on what resistors are used. Here, an input of 0.25 ma can give an output that is 40 times greater, in other words, 10 ma. A common emitter circuit can also amplify voltage at the same time as current, which is "power amplification."

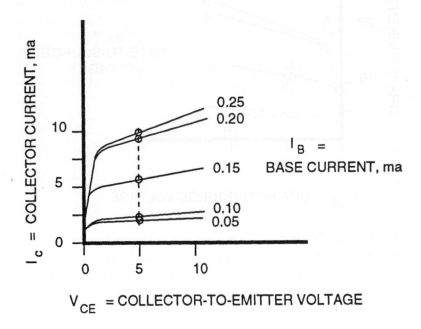

The vertical dashed line could be used to calculate a more intuitive S-shaped curve of collector current versus base current, just like the dashed line on page 269 was used to calculate the 100-volt curve on page 270, but we will not bother to do that here. Also, a load line could be drawn, as on the previous page.

GLOSSARY

This section includes some less-common meanings of words and phrases. The items included here can be confusing, especially to people who are not familiar with electronics, but also to people for whom English is a second language.

(For further explanations, see also the Index or Contents.)

ANALOG

A system that behaves similarly to the way another system behaves, even though they are not exactly the same. An example would be that some water pipe systems can be analogs of some electronic systems.

Also, a method for encoding information in a manner that is proportional to the original. In other words, sound waves can be represented by electrical signals so that, the louder the sound gets, the stronger the electric signal is, in direct proportion to each other. Another meaning in electronics, with the word being used as an adjective again, is linear amplification.

ARC, ELECTRIC

A *long-term* avalanche conduction in a material which was previously an insulator. It is caused by an electric field that is higher than the "dielectric strength" of the insulator. It is limited in current, only by *some other* resistance (not its own resistance), and it can have great power output as heat. It has a "negative incremental resistance," which is the reason why its current is only limited by some external resistance value. (Compare to "SPARK" on page 283.)

BALANCED LINE
Two conductors, making up part of a circuit, neither of which is grounded. Electromagnetic waves hitting the pair will tend to induce the same voltages in each, which can be cancelled out by suitable circuitry at the ends of the wires.

BALUN
A component, usually a transformer, which can have a "balanced line" input (no grounds) but an unbalanced output (having one conductor be grounded). Often it also serves to match impedances, especially the "characteristic impedances" that can lead to reflections if they are not matched.

BRIDGE
A diamond-shaped group of wires, usually with current coming into the top and bottom and going out of the front and back corners. It can involve resistors, rheostats, rectifying diodes, or other components. It is often used to detect or measure small amounts of voltage.

CERAMIC
A nonmetallic material, usually an oxide, made from powder particles that have been "sintered" at high temperature, to merge them into a single unit. Sintering

involves getting the particles close together, sometimes by "dry pressing" or wet "tape casting," and then heating them to a temperature that is roughly about 1,000°C but is below the melting point of the main material. (However, about 1% of the total material composition almost always does melt, leading to "liquid phase sintering," in addition to "solid state diffusion" sintering.)

CHARACTERISTIC IMPEDANCE, Z_0

The impedance of an infinitely long transmission line. If wire pairs with *different* characteristic impedances are connected together, much of the electrical energy will usually be reflected back into one pair, instead of transmitting all of it into the other pair of wires. Such reflections can be prevented with a "balun" or a "matching stub," but it is more desirable to design the whole circuit so that the impedances are matched.

CHASSIS

A mechanical support for components, usually made of aluminum metal. It can provide a convenient ground connection, and it can also act to shield the components from EMI.

COAX

A wire that carries electrical signals, inside a tube of insulating material (usually plastic such as polyethylene), which is itself inside a conductive metal tube (usually wire mesh or screen). This tends to prevent EMI. It is more effective than a twisted pair, but more expensive. It is often used because of its special "transmission line" properties for high frequencies.

COLD JOINT

A soldered joint which looks good in visual inspection but actually is not electrically conductive. Usually it was made at too low a temperature to evaporate the rosin flux out from between the solder and the copper wire.

CONTACT BOUNCE

Electrical contacts in relays and switches often bounce several times when making contact ("closing"), instead of just smoothly closing once. This can cause the electricity to start and then stop quickly, which can enter digital data the wrong number of times. A "debouncer" circuit (such as a Schmitt trigger) can prevent the multiplicity of startings.

DIELECTRIC STRENGTH

An electric field, above which a material will probably be destroyed by the formation of a spark ("avalanche conduction").

DIGITAL

A method for taking a short-time sample of a voltage, and assigning a coded signal to it. This code consists of a group of even shorter-time pulses ("square waves"), which represent the original voltage at that particular time when the sample was sensed. This process is called A-to-D (analog to digital) encoding.

At the other end of the signal path, the opposite process takes place, D-to-A decoding, which converts the signal back to a voltage proportional to the original one. The equipment that does both of these can be called a "codec," which is short for coder-decoder.

DVD

Letters that previously stood for words such as "disc," but now do not stand for three words. Taken together, they refer to a standard for digitally encoding music and video, and storing it in a CD-configured plastic medium. (Note that "disc" is spelled with a "c" when referring to music storage, but with a "k" if used for other purposes such as computer hard drive memories.)

EMI

Electromagnetic interference, similar to RFI, but sometimes involving a lower frequency such as 60 Hz. It can induce spurious signals in unshielded wires, causing errors in computers or in telephone transmissions of bank data, etc.

EMF

In older literature, "electromotive force," which is voltage. In modern usage, it means "electromagnetic fields," which includes radio waves, magnetism from power transformers, and similar fields. Some environmental scientists say that it can be damaging to human beings, but others say it is harmless.

GROUND LOOP

A system in which two components are connected to the ground imperfectly, so that current in one part induces an undesirable voltage in the other part. This can lead to oscillations or mistaken measurements. Solutions to these problems can involve using much lower-resistance ground wires, or connecting both components to the same ground through a single wire. Sometimes it can be prevented by making sure that there is no circular path for electricity flow within the grounding wires, simply by cutting one of the connections that link the two components together, via an imperfect ground path.

GRABBER

A word sometimes used in electronics catalogs, when describing various alternative types of cables for attaching instruments. A grabber is an adjustable connector at the end of a wire, often used with oscilloscopes, which can be squeezed by hand. When this is done, a metal barb comes out and can be hooked onto a wire that is being tested. Releasing it allows a spring to ensure good contact.

GUARD

This can be an ordinary grounded "shield" conductor. However, it usually refers to a third conductor, which is neither grounded nor attached to the input signal. It tends to prevent both EMI and also stray dc interference.

HOOK

A word sometimes used in electronics catalogs, when describing various alternative types of cables for attaching instruments. A hook is an adjustable connector at the end of a wire, often used with oscilloscopes, which can be squeezed by hand. When this is done, a metal barb comes out, and it can be hooked onto a wire that is being tested. Releasing it allows a spring to ensure good contact.

IMPEDANCE

The total resistance to the flow of ac electricity that is made up from the ordinary dc resistance, in addition to the current-limiting effect of the ac "reactance." (They are not simply added together.) For dc it is the same as resistance, but for ac of high frequencies it can be very different.

INDUCTIVE KICK

A pulse of voltage, usually with very short duration and quite high voltage, which is generated by suddenly stopping the current that is going through a highly inductive coil of wire. It can damage other equipment, sometimes including things that are far away but are attached to the same wires. In well designed inductive equipment, it can be "snubbed" down to low voltage, by using a capacitor or diode.

JUMPER

A short, removable section of wire or "lead." This is sometimes called a "jumper cable," and the definition is the same as for the word "patchcord."

LEAD

In electronics, this word has two meanings. One, in which it is pronounced "led," refers to the metal, which is used in solder (typically 40% lead, 60% tin).
The other meaning is a wire, or any piece of metal that carries electricity. This is sometimes a flat peice of metal, rather than having a round cross section.

LUG
A short wire, or end of a wire, which is large and therefore easy to solder to or make screw contact to, compared to the thin wire itself.

MULTIPLEXER (MUX) and CODEC
With multiplexing, many different signals can be sent via the *same* transmission medium, but they can be *separated* somehow at the receiving end. An example is the common situation where several radio stations all transmit through the same atmosphere, and a certain receiver can tune to just one of those stations, but another receiver can receive a different station by tuning to a different wavelength. This is "wavelength division multiplexing," or WDM. (Of course, the wavelength is related to the *frequency* as shown at the bottom of page 212, so radio could be called *FDM*, but we usually don't say it that way.)

Another method for sending several signals over the same wire or optical fiber is to have each signal occupy a dedicated *time* slot. This is often done with digital transmission, which is called "time division multiplexing," or TDM.

Still another way to separate various signals is to have them each be transmitted via a *different wire,* all within a cable, and this is "space division." It is commonly used at both ends of a telephone system, while WDM or TDM are often used in the intermediate "trunk line" between two cities.

A complex electronic apparatus that accepts various incoming signals and *mu*ltiplexes them into a single transmission medium is called a MUX. If it changes analog signals into digital signals that are pulse *code* modulated (PCM, page 213), and it also can *decode* signals coming back in the opposite direction, then it is called a CODEC, meaning *co*der and *dec*oder. (See also the explanation of the word "modem" at the top of page 209.) The reader is likely to hear these terms mentioned quite a lot in the near future.

NEGATIVE RESISTANCE

A confusing term, since true negative resistance does not exist. (If it did, energy could be generated from nothing.) A better description of the phenomenon is "negative *differential* resistance," in which a resistance decreases when the current increases. Examples are neon bulbs and PNPN diodes, just after the current has increased sufficiently to "turn them on."

OHMMETER

An instrument that can be used to measure resistance in ohms. It contains a voltage source and a means for measuring current, from which resistance can easily be derived.

OHM-METER

A unit of resistivity, defined in terms of the resistance of a cube, one meter of length on each side. A similar term is ohm-centimeter, measured on a smaller cube.

PATCH

A short, removable section of wire or "lead." This is sometimes called a "patchcord," and the definition is the same as that of the word "jumper."

POTENTIAL DIFFERENCE

Another phrase meaning voltage. Sometimes just the word "potential" is used, also meaning voltage.

POTENTIOMETER

Unfortunately, a word with two alternative meanings. One refers to a resistor whose value in ohms can easily be changed, usually by rotating a shaft or knob.

The other meaning involves a variable resistor, as above, but it is attached in such a way that an output *voltage* can be varied, not just the *resistance*.

RELAXOR

A chemical compound made up from elements such as titanium, niobium, lead, magnesium, and oxygen, which can have an extremely high dielectric constant (sometimes more than 20,000). These can also be piezoelectric, with very good properties.

RFI

Radio frequency (higher than 20 kilohertz) interference. Radio waves can accidently induce electricity in unshielded wires, interfering with the signal in the wire. An example is that strong *radio* signals can sometimes be heard in a stereo system that is only supposed to be playing music from a CD or DVD.

RHEOSTAT

A variable resistor, which is the first-mentioned meaning of "potentiometer," listed above. Usually it is a large component that can handle high currents and be cooled by convective air flow.

SCATTERING

A phenomenon by which imperfections in electrically conductive materials such as metals can raise the resistance. These imperfections might be permanent, as with chemical impurities, or temporary, as with heat-induced movements of the atoms.

SHUNT

A wire which goes around another conductor, so that they are both connected in parallel.

SNUBBER

A capacitor plus resistor (usually both of small values), which is arranged to catch "inductive kick" voltage pulses and harmlessly short circuit them. Another way to do this is with a rectifying diode. Still another way is with a Zener diode or argon gas tube.

SPARK

Avalanche conduction in a material (or vacuum) which was previously an insulator. It is caused by an electric field that is higher than the "dielectric strength" of the insulator. It is limited in current, either by *its own* resistance, or by using up the available coulombs of charge, after which the spark dies out. It is different from an "arc," which is limited only by *some other* resistance, and which can have great power output as heat.

SQUARE WAVE
An electric signal that is usually rectangular, not really "square." The voltage rises quickly, then stays constant, and then goes to zero quickly. This is the shape of most pulses used for digital coding.

THICK FILM
A layer of conductive material, usually about 25 micrometers (one thousandth of an inch) thick, mounted permanently on a thicker supporting "substrate," made out of ceramic. Conductors, resistors, capacitors, and other components can be made from thick films. They are usually made by screen printing powdered metal, plus glass powder, plus organic glue onto the substrate, and then firing these at about 850°C. The glue burns away, and the metal sinters to become continuous. The result is a "cermet," like the thick film resistor mentioned near the bottom of page 162.

THIN FILM
A layer of conductive material, usually about *one* micrometer thick, mounted permanently on a thicker supporting "substrate" (usually ceramic, but sometimes silicon, etc.). Conductors, resistors, capacitors, and other components can be made from thin films. They are usually made by evaporation or sputtering in a vacuum chamber. The quality is usually better than that of thick films, but the cost is usually higher.

VALVE
The device that is called a "vacuum tube" or just a "tube" in the U.S.A. is usually called a "valve" in England.

INDEX

(See also Glossary, page 273.)

B

F

P

X

Y

Z

GREEK LETTER SUPPLEMENT

(Lower case letters, unless otherwise
indicated)

Printed and bound by CPI Group (UK) Ltd, Croydon, CR0 4YY

03/10/2024

01040433-0014